Move Slowly and Build Bridges

# Move Slowly and Build Bridges

## *Mastodon, the Fediverse, and the Struggle for Democratic Social Media*

ROBERT W. GEHL

OXFORD
UNIVERSITY PRESS

Oxford University Press is a department of the University of Oxford.
It furthers the University's objective of excellence in research, scholarship,
and education by publishing worldwide. Oxford is a registered trade mark of
Oxford University Press in the UK and in certain other countries.

Published in the United States of America by Oxford University Press
198 Madison Avenue, New York, NY 10016, United States of America.



Library of Congress Cataloging-in-Publication Data
Names: Gehl, Robert W. author
Title: Move slowly and build bridges : Mastodon, the fediverse, and
the struggle for democratic social media / Robert W. Gehl.
Description: New York, NY, United States of America:
Oxford University Press, 2025. | Includes bibliographical references and index.
Identifiers: LCCN 2025006518 (print) | LCCN 2025006519 (ebook) |
ISBN 9780197776681 (paperback) | ISBN 9780197776674 (hardback)
Subjects: LCSH: Social media—Moral and ethical aspects | Open source
software—Social aspects | Freedom of information | Mastodon
Classification: LCC TK5105.878 b.G47 2025 (print) |
LCC TK5105.878 b.G47 2025 (ebook) | DDC 302.23/1—dc23/eng/20250416
LC record available at https://lccn.loc.gov/2025006518
LC ebook record available at https://lccn.loc.gov/2025006519

DOI: 10.1093/9780197776711.001.0001

Paperback printed by Marquis Book Printing, Canada
Hardback printed by Bridgeport National Bindery, Inc., United States of America

The manufacturer's authorized representative in the EU for product safety is
Oxford University Press España S.A., Parque Empresarial San Fernando de Henares,
Avenida de Castilla, 2 – 28830 Madrid (www.oup.es/en or product.safety@oup.com).
OUP España S.A. also acts as importer into Spain of products made by the manufacturer.

*This book is dedicated to my mom, Pamela. Only a mother would shelve her son's books next to Shakespeare in her home library. It always brings a smile to my face when I see this.*

*Thank you, mom!*

# Contents

# Acknowledgments

I began writing this book in earnest in early 2021. Much has changed since then. Professionally, I watched as the study of alternative social media shifted from a somewhat obscure corner of research to a major area of study, with conferences, a growing body of journal articles, and new scholars doing cutting-edge work. I also moved institutions, joining the Department of Communication and Media Studies at York University. This also meant moving countries, leaving the USA for Canada. These professional changes were driven in part by events documented in this book.

I thus owe gratitude to the people of the fediverse in more ways than one. First, thank you to all who agreed to be interviewed or quoted in this book. I want to particularly thank Artist Marcia X not only for participating in an interview, but also for agreeing to license her art—their image "Untitled 7" is the image on the cover of this book. But beyond thanking members of the fediverse for their contributions to this book, I need to thank them for being politically engaged and diagnosing many of the disruptions we now face. The political events of the past few years were not surprising to longtime developers, admins, moderators, and members of the fediverse. The pages that follow, I hope, will reveal why.

Outside of the fediverse, I have benefited from the insights, ideas, and support of academic colleagues around the world. These include, in no particular order: Maria Bakardjieva, Gordon Gow, Roel Roscam Abbing, Diana Zulli, Nate Tkacz, Carlos Cámara-Menoyo, Iain Emsley, Alex Martin, Jessa Lingel, Tero Karppi, Nathalie Van Raemdonck, Zenna Elfin, and Jamie Theophilos. I want to especially thank the Association of Internet Researchers for having the wisdom to run its own social media and allowing me.

The support of York University has been invaluable. This includes my excellent and supportive new colleagues in Communication and Media Studies—I want to especially thank Ganaele Langlois, Natalie Coutler, Kelly Bergstrom, and Nick Taylor. I also want to thank the Canada First Research Excellence fund. Thanks also to the students and faculty who have invited me (or simply endured me) giving lectures on the fediverse. I particularly want to thank a York student, Cassidy Green, for sitting with me to figure out how the ethics of care could be a lens to look at the fediverse.

The folks at Oxford University Press have been wonderful to work with. The book is better for the editorial guidance of Norm Hirschy. I am indebted to Nick Couldry and Josh Braun who provided the anonymous reviews of this manuscript and then later revealed their identities and shared more criticism and feedback.

As always, my love goes out to friends and family who have supported me. Sean, Cynthia, and Everett—thanks for visiting us and seeing Lake Michigan at sunset. Ry, Brian, Mony, and Dan: you keep me in check. None of this would be possible without my partner Jess and my puppy buddy Teddy. Aggie also is an excellent therapist. Throughout the turmoil you have kept me hopeful that a better world is possible.

# Introduction

## On Alternative Social Media

Much of my academic career has been dedicated to answering one question: Where do people go online to socialize if they do not like how corporate social media—Facebook, Instagram, Twitter/X, YouTube, and TikTok—are run? This book is about one specific answer: they go to alternative, community-run social media systems, such as Mastodon and the fediverse.

I've been researching and writing about alternative social media since I was a PhD student in the late 2000s.[1] As I've defined it in my research, alternative social media are not controlled by corporations, but by ordinary people.[2] Unlike Facebook or Twitter/X, alternatives are not run for profit. In contrast to corporate systems, alternative social media tend to be free and open-source software (FOSS), meaning that anyone can study, download, install, and run these systems. This allows people to have greater insight into their technical designs than they get from the corporate counterparts.[3]

In addition, the people who develop and run them tend to be open about their motivations. Over my career, I've interviewed activists, technologists, and everyday people who have quit corporate social media and opted instead to build and use community-owned systems.[4] I've also observed as millions of people have adopted alternative systems instead of corporate ones. They do so for many reasons, ranging from concerns over surveillance to frustrations with moderation to anger over algorithmic manipulation.

When I say "alternative," I'm placing alternative social media in a long history of alternative media. For as long as there have been media, there have been alternatives to the mainstream forms, such as underground newspapers sold on street corners, pirate television stations made by everyday people, and what media studies scholar Clemencia Rodríguez calls "citizens' media"—low-power radio, pamphlets, and street orators.[5] As researchers have shown, these systems are alternative not just in the sense of the content they publish but also in terms of who controls them—activists and everyday

*Move Slowly and Build Bridges*. Robert W. Gehl, Oxford University Press. © Oxford University Press 2025.
DOI: 10.1093/9780197776711.003.0001

people whose concerns are radically different from those of the operators of mainstream media.[6] As Rodríguez puts it, the people who make and run alternative media seek to "alter the social world in which they operate," providing the rest of us with a way out of corporate or state-controlled ways of thinking.[7] Much as these older media do, alternative social media challenge the concentration of media power in the hands of Google, Meta, and Twitter/X.[8]

In addition to revealing details of how social media works, alternative social media reveal details about how people govern themselves online, which is a growing area of scholarly concern.[9] Instead of ceding decisions about content moderation and access to corporations, the people building alternative social media systems are shifting "our use of social platforms towards ones that communities own and govern," as digital media scholar Ethan Zuckerman puts it. Drawing on the sociologist Robert Putnam, Zuckerman argues that alternative social media can bolster the practices of democracy. "We should make this shift because participating in social networks we ourselves govern could turn online spaces into Putnam's 'schools for democracy.' "[10] Instead of relying on corporations to manage one of the most important things we humans do—communicate—the activists, technologists, and everyday people making alternatives want to bring media under our control. They want to make social media *ours*.

I'm deeply compelled by the effort to make social media ours. My journey into alternative social media grew out of my own increasing frustration with corporate social media, particularly Facebook. I spent my early years writing a critique of corporate social media, culminating in my first book, *Reverse Engineering Social Media*, in 2014.[11] Drawing mainly on Marxian critiques of communication, I faulted Facebook and Twitter for converting human sociality into commodities.[12]

Critique of corporate social media is now a major concern of fields such as media studies and software studies.[13] However, unlike many of my peers, I argue for something more than mere critique. In my view, despite years of criticism leveled at Facebook (now Meta), Twitter (now X), or YouTube, their commodification and manipulation of sociality continue apace. This is why I argue for continuous critical reverse engineering of corporate social media, where we use the critical knowledge we gain from analyzing corporate social media "to produce new artifacts that simultaneously improve upon the old and yet also bear a relation to the old."[14] The "critical" part of this is attempting to build a new system that challenges the concentration of power found in the old: ethical, community-run alternative social media. I concluded *Reverse Engineering* with a "Manifesto for Socialized Media," a call for social media that is controlled by ordinary people, avoids centralized control, and eschews what we now call "surveillance capitalism" in favor of noncapitalist approaches. That manifesto set me on the path I am on now, and one result is this book.

My years of study of alternative social media have revealed many practices that we simply cannot see in corporate social media, ranging from how the software works—including the algorithms—to how people do the work of moderating them, how they move our data around, and how members can actually be in control of their data.[15] Because alternatives reveal much more to researchers than their corporate counterparts, I argue that studying alternative social media adds new perspectives to fields of study such as social media studies, software studies, and Internet studies. This book, I hope, proves this point.

Of course, there is one area where alternative social media does not compare favorably to corporate social media. If we compare corporate social media to alternatives, the populations are almost incommensurate. For example, at its height, the Twitter-like GNU social had tens of thousands of users.[16] Corporate social media have billions of users. Mastodon, the focus of this book, counts only about 10 million accounts.[17]

Even that reveals something important, however: the activists, technologists, and ordinary people making alternative social media aren't giving up just because these systems aren't as popular as corporate social media. People who make alternative social media do so in large part due to continued frustrations with corporate social media. Alternative social media activists continue to believe something better is possible.

Mastodon and the fediverse are the most recent—and in my view, most important—attempts to make something better than corporate social media.

## Joining Mastodon

I joined Mastodon in 2017 because of this long-standing research interest in alternative social media. Founded in 2016, Mastodon is a FOSS microblogging site. In this way, Mastodon was similar to many other existing Twitter-like alternative social media systems I was familiar with, such as GNU social, rstat.us, Twister, and TalkOpen.[18]

When I joined Mastodon, I was leading a research project, the Social Media Alternatives Project, at the University of Utah, where several graduate students and I were building an archive of documents about alternative social media.[19] I had been studying alternatives for nearly a decade at that point, and my approach involved finding alternatives, signing up, trying them out, and studying their cultural norms and technical infrastructures. This exploration included signing up for systems like Twitter-like alternatives, as well as others: diaspora*, Lorea, and various social media sites that were hosted exclusively on the dark web.[20]

I didn't think much of joining Mastodon in 2017, at least at first. It was one of many alternatives I had examined. But it did have all the features I could want: Mastodon allows members to do Twitter-like things, such as follow, be followed, use hashtags, make short posts, favorite other posts, and boost them (akin to a retweet). It had a clean interface. It was FOSS. Importantly, it was federated.

One simple way to explain federation is to consider email. If a friend of mine has a Gmail account, and another has an Outlook account, and I have a Protonmail account, our accounts exist on different services. One is owned by Google, another by Microsoft, and the other by Proton. Even though these are three different companies, my friends and I can communicate via email, which operates using a shared technical protocol. These different email servers—or "instances," to use fediverse parlance—are connected, or federated, even though each is independent of the other. And much like Mastodon, there are FOSS email hosting systems available, allowing people with the technical know-how to run their own email servers and yet communicate with people on corporate servers.[21]

Federated social media is a radical departure from corporate social media. While big companies like Google and Microsoft produce email servers that can intercommunicate with hundreds of other providers, in the world of social media, cross-company communication is largely unheard of. A Facebook user can't send a message to a Twitter/X user, for example, nor can a TikTok user comment on a video hosted by YouTube. This is because Facebook, Twitter/X, Instagram, and TikTok are not federated systems, but centralized systems, seeking to lock people into them.[22] Indeed, their success hinges in part on keeping people on their platforms through a combination of a large userbase and attention-grabbing, algorithm-generated feeds. Cross-corporation communication would undermine this system.

Beyond cross-site communication, there is something even more radical to consider when we compare federated social media to Facebook or Twitter/X. Federated social media, like email, can be hosted by individuals or small groups of people. In contrast to Facebook, where people can sign up for accounts on a corporate-owned server, federated social media allow people to run their own servers, allowing friends, family, or strangers to sign up and socialize. Mastodon is exemplary in this regard—while it is now a network that spans much of the globe, it comprises tens of thousands of small communities averaging 500 accounts in size.[23] Governance among 500 people is a far different matter than corporate social media's attempt to govern billions of people globally.[24]

## The Fediverse and Noncentralization

There is something even more radical than Mastodon's instance-to-instance federation, however. Mastodon is part of an even larger network of social media sites called "the fediverse." On the fediverse, Mastodon is joined by Pixelfed, an Instagram-like photo sharing service; Friendica, which is similar to Facebook; PeerTube, which is a YouTube alternative; Lemmy, a Reddit-like system; and Plume for blogging. There are also other microblogging systems, such as Misskey, Pleroma, Firefish, and GNU social. *Each* of these software packages allows anyone with the technical know-how to install them on an instance and let people sign up for accounts.

Besides being alternatives to corporate social media, these sites also offer a unique feature: anyone with an account on one of these systems can potentially follow others—*even if they're not on the same system*. From my account on a Mastodon instance, for example, I could follow a PeerTube member and see their posts, and vice versa. A Friendica member can follow a Plume blog and comment on it. Pixelfed images can appear on a Lemmy instance. A Friendica member can follow a GNU social member who follows a Mastodon member. This is also very much unlike corporate social media. A Facebook user cannot follow a Twitter/X user—or even an Instagram user, despite Facebook and Instagram being part of the same company! People who use Facebook, Instagram, Twitter/X, or TikTok are locked in. In contrast, on the fediverse, people can follow people across different types of social media. People are not locked into whichever of these they choose—they can access others from across a range of different social media applications.

The result is a heterogeneous mix of potentially interconnected social media services spanning the globe. The best estimate for the total number of fediverse instances is 27,000, with around 10–14 million accounts across the network.[25]

This broader network of heterogeneous social media systems, the fediverse, is possible because of a shared technical protocol called ActivityPub. This protocol allows any instance that obeys its rules to connect to others. Much like email—which runs on a standard protocol—ActivityPub structures the connections between participating fediverse instances. ActivityPub is also an open standard, developed in 2018 by the World Wide Web Consortium, the same standards group that helps make Web technologies work (more on its development in a later chapter).[26]

Taken as a whole, the ActivityPub-based fediverse is a direct challenge to the centralization found in corporate social media. We often call this approach "decentralization," but as I will argue, it's more accurate to call it "noncentralization," a term I borrow from political philosophy.[27] "Decentralization" implies that there was once a center, such as a corporate entity, or a single server, that is broken up into multiple, smaller centers—and it also implies that what gets broken up can be recentralized later.[28] Noncentralization, by contrast, means designing a system to actively avoid any center of power emerging from the outset. While alternative social media developers often speak of decentralizing social media, in practice I find that they attempt to build systems that avoid centers in the first place. Mastodon and the rest of the ActivityPub-enabled fediverse is, as I will show, not just one site, but many distinct instances, each with their own rules and cultures.[29] Rather than starting with a central, powerful site and breaking up, the ActivityPub fediverse tends to start with small social media instances that network with one another into a large network.

If the ActivityPub-enabled fediverse is a network of tens of thousands of instances, many of them community-run, what holds it together? Arguably, the answer is technical: the ActivityPub protocol itself provides the infrastructural connection. But the more I have observed the fediverse, the more I realized there is more to it than technical protocols. The administrators and moderators of fediverse servers also make social choices about which other servers to connect with. This is the fediverse's most radical departure from corporate social media. What emerges, I will argue in this book, is a "covenantal fediverse," a network where ethically aligned communities band together in a larger network. In my view, while noncentralization and community governance are important departures from corporate social media, what sets Mastodon (and much of the fediverse) apart the most is the development of an ethical covenant as a social protocol that bands the heterogeneous network together.

I joined Mastodon in 2017, and very quickly, it and the broader fediverse appeared to me to be closest to the ideal alternative social media system I had laid out in my "Manifesto for Socialized Media." It was then that I began the research that would result in this book.

## Focusing on Mastodon

Mastodon is a Twitter alternative. Pixefed is an Instagram alternative. PeerTube is a YouTube alternative. Each allows for hundreds of distinct installations, all potentially connected via the open technical protocol ActivityPub. While this variety of social media is exciting to see, all of these options present a challenge to me as an author. Attempting to represent all ActivityPub-based social media would indeed make this book too complex. For clarity's sake, this book focuses on Mastodon.

In part, I'm doing so because at the time of this writing it is one of the most popular fediverse systems. While counting membership is challenging in a noncentralized network, Mastodon currently has anywhere from 7 million to 10 million members, growing particularly rapidly after Musk's purchase of Twitter in late 2022.[30] In addition, there's more to Mastodon than size and popularity. Its history is very important to the fediverse itself. Mastodon played a pivotal role in the development of the ActivityPub protocol—in fact, the ActivityPub-powered fediverse itself might not even exist if it weren't for Mastodon. Most importantly, studying Mastodon reveals five key themes—more on that below.

First, however, let me say something about methods. To study Mastodon, I deployed the approaches I was trained in, mixing participant observation, software studies, and interviews to unpack Mastodon's complexity. (You can read more about my methods in the Appendix.) My participant observation has been influenced by digital ethnographers—I have sought to immerse myself in the daily practices of fediverse members.[31] I began with an account on a Mastodon instance in 2017, and I ran my own test instance to try my hand at self-hosting. I studied Mastodon's underlying code and analyzed how the code shaped (and was shaped by) social interactions. I gathered journalistic coverage of Mastodon. I kept a research journal throughout this period using Zettlr, a FOSS notebook that allows for tagging and connecting ideas. Eventually, I was asked to run a Mastodon instance on behalf of the Association of Internet Researchers, which gave me more insights—specifically into how moderation and administration work. Throughout these observations, I would note key concerns, crises, and pleasures that emerged on Mastodon and across the fediverse. These helped reveal the themes of the book.

As for software studies, as I discuss in a previous book, engaging with software is just as intense as interviewing people.[32] Hosting a Mastodon instance, as I have for several years, requires constant attention: finding a location on the network, watching the CPU load of the server, watching for updates, reading manuals, and selecting options. And this is just running the software—to study fediverse software effectively, one also must examine para-texts, such as Github issues, developer blogs, email lists, and public debates over features. These approaches must then be mapped back to the cultural practices of the use of software. As software studies scholar Matthew Fuller puts it, software studies involved "a rich seam of conjunctions in which the speed and rationality, or slowness and irrationality, of computation meets with its ostensible outside (users, culture, aesthetics) but is not epistemically subordinated by it."[33]

This book, however, is not about my experience. I have attempted to center the voices of developers, admins, moderators, and everyday members of the fediverse, predominantly through qualitative interviews.[34] Most interviews were conducted via video chat (on a self-hosted Nextcloud server) and lasted over an hour. A few were conducted via email, Signal messaging, or direct message, depending upon the preferences of the interviewee. All interviewees were offered a copy of my transcript of our conversation, and some asked to see draft chapters. I granted those requests and was often rewarded with invaluable feedback resulting in more themes for this book. In addition to interviews, I have also drawn on public sources from fediverse members, such as blog posts, as well as more private sources, such as fediverse posts and Patreon members-only posts, archived in Zotero, a FOSS bibliography application. These latter sources are used with the explicit consent of their authors.

Overall, this is a work of Internet studies. Scholars in Internet studies now analyze technical details, such as algorithms, conduct "walkthroughs" of the interfaces of applications, and examine digital infrastructures. However, they also place these things in relation to the compelling and often wild range of practices users coax out of digital media.[35] I have sought to extend this sort of analysis to the Mastodon and the fediverse, considering it as embedded in broader sociotechnical contexts, while still treating it as a kind of social world unto itself.[36]

## Themes of the Book

The history, technologies, and cultural practices of Mastodon open up the five themes I will explore in this book. First, studying Mastodon reveals how people go about making alternative social media and how alternatives relate to corporate social media. Second, a major theme centers activists who are struggling over implementing ethical, community-centric governance in social media and creating what I will call the covenantal fediverse. Related to this, a third theme discusses how the covenantal fediverse has in part rejected dogmatic free speech absolutism, and with it an overemphasis on the individual social media user. A fourth theme is about how marginalized people—particularly queer, trans, and Black—have struggled to improve the covenantal fediverse, even at great costs to themselves. Finally, the fifth theme explores how technical and social innovations can happen in alternative social media despite the lack of involvement of large corporate social media platforms.

### Making Alternatives

Mastodon and the rest of the fediverse are alternative social media. By definition, something that is "alternative" has a relationship to something else, usually called "mainstream" or "standard." In the case of the fediverse, that relationship is to the corporate social media sites that have dominated online sociality for the past two decades, particularly Twitter/X, Facebook, and Instagram. To be legible as an alternative to Twitter, for example, Mastodon must have Twitter-like features. However, it also must attempt to prevent Twitter-like problems—this is the critical reverse-engineering approach. Attempting to avoid Twitter's problems is what makes Mastodon an "alternative."

I will not reduce "alternative" to a mere choice—it's not like a consumer choice between, say, Coke or Pepsi, Twitter or Mastodon. Instead of being a mere choice, alternative social media involves different social structures than their corporate counterparts. Most often this involves giving power to different sorts of people. The corporate model places authority in centralized groups (CEOs, owners, boards of directors). Alternatives tend to privilege activists, technologists, admins, moderators, and creative people as the authorities. And, in theory, ordinary participants in the network can become more active in Mastodon development, moderation, and administration.

I will highlight this distinction by referring to fediverse participants as "members" instead of "users," since "users often have limited input in developing and implementing tools they use, while members are part of a larger decision-making collective where they can exert agency."[37] Instead of merely signing up for an account and using a service, alternative social media practitioners often invite ordinary people to become more involved in many aspects of operating and governing social media. This process will be discussed in Chapters 1, 2, and 3.

A central concern here, then, is, Will this alternative be viable? Can the fediverse withstand all the pressures it will face, such as attracting enough people, dealing with difficult moderation problems, and paying the bills (since running an instance is not free)? And can the fediverse survive challenges from corporate social media? For example, the world's largest social media corporation, Meta, now runs an ActivityPub-enabled microblogging system called Threads that can federate with Mastodon and other fediverse services. Will this undermine the status of the fediverse as a noncorporate social media alternative? I will discuss the viability of alternative social media in several chapters, especially Chapters 1, 2, 6, 7, and 8.

## The Covenantal Fediverse

Critical reverse engineering includes a normative component: the goal is to make something better than the original.[38] What is "better" can vary quite a bit. In my research, I've studied alternatives that center protections for free speech, seeking to avoid censorship. Others center privacy concerns, looking to avoid the surveillance practices of corporate social media. Still others tout their alternative technological structures, attempting to create noncentralized or even peer-to-peer network topologies. Often, these concerns overlap.

The fediverse includes people who have these concerns. However, one key segment of the fediverse has created a unique alternative social media structure—a particular set of ethical practices I call the covenantal fediverse. Many fediverse instances openly proclaim their ethical values and use those proclamations not only for internal content moderation but also for maintaining or severing instance-to-instance relations.

I borrow the concept of the covenant from political philosophers who have examined federalist political structures, such as cantons in Switzerland or the Haudenosaunee Confederacy in North America.[39] A covenant is "a morally informed agreement or pact between people or parties having an independent and sufficiently equal status; it is based on voluntary consent and established by mutual oaths or promises."[40] Covenants establish a baseline of ethical values. Autonomous groups who agree to those ethical values can band together while still maintaining their respective sovereignty.

In the covenantal fediverse, federation is not just a technical practice—it's a social and political one.[41] It's not merely achieved by linking instances through the ActivityPub technical protocol. The covenantal fediverse comprises a noncentralized network of small, community-run instances who bond through shared ethical values.[42] Not unlike the United Federation of Planets of *Star Trek* fame, the fediverse brings together many different communities, each with their distinctive cultures and codes of conduct, into one larger community. Fediverse members are simultaneously members of instances as well as of a broader network of instances.[43] The fediverse thus highlights another meaning of the term "protocol"—rules of etiquette and diplomacy.

Creating and maintaining the covenantal fediverse is difficult and contentious work. The governance practices and cultures of covenantal fediverse, including related controversies and struggles, will be discussed throughout this book, but especially in Chapters 4 and 5, and the Conclusion.

## Denying the Dogmas of Free Speech Absolutism and Individualism

The covenantal approach to federation emphasizes groups, typically found at the instance level. Due to this, instead of freedom of speech, covenantal fediverse members tend to highlight freedom of association, the ability to choose with whom they connect, typically through signals such as codes of conduct. Codes of conduct often do restrict speech, banning racist or transphobic posts. However, the result of this is not a limit on free speech, but rather a limit on what philosopher Herbert Marcuse calls "repressive tolerance," where dominant ideas can drown out those of marginalized people.[44] Marginalized people and their allies on the fediverse have struggled to make a social networking system where queer, trans, and people of color can express themselves because they are not overwhelmed with hate speech.

As two fediverse scholar-practitioners write, strong moderation on the fediverse challenges dogmatic visions of open communication. "What is unique about the fediverse is both technical and cultural acknowledgement that openness has its limits, and is itself open to wide-ranging interpretations dependent on context, which are not fixed in time. This is a fundamentally new point of departure for reimagining social media today."[45] Whether this is successful is a matter of debate, as I will show particularly in Chapter 5.

Related to this, the covenantal fediverse also centers groups over individuals. Rather than seeing the individual as the atomic unit of social media, the unit who speaks to a listening world, the covenantal fediverse centers instances-as-communities. Communities make choices about the freedom to associate. This is another reason I will use "member" instead of "user" to refer to fediverse participants. As I show in Chapter 4, codes of conduct are especially important here. I will focus predominantly on Mastodon instances as a key unit of the covenantal fediverse. The act of starting an instance is both a technical and a political act: an instance admin must choose how and where to host the service, set policies, decide who should be allowed to sign up, and decide how long connections between instances should be maintained. Even if an admin shirks these things and says "anything goes," that's still a political decision.[46]

This is not to say I will ignore the experience of individual Mastodon and fediverse members, but that I will consider the individual's experience with the fediverse in relation to the instance level. Much like starting an instance is a political choice, selecting an instance to join is a political choice—and as a result, having to make such a choice to engage in social media can be a major barrier to fediverse adoption. Chapter 2 discusses the challenges facing individuals who must select an instance to join.

## Marginalized Folks Move to the Center

That the covenantal fediverse enjoys the active contributions of queer and trans people is not surprising. Recent Internet histories have revealed the central role queer and trans folks have played in digital technologies. As gender and sexuality studies scholar Avery Dame-Griff notes, "each step of the way in the history of computer communication . . . trans people have been there."[47] Queer and trans people's contributions to the fediverse have not only included safety and moderation tools but also contributions to the ActivityPub protocol itself. These offerings are predominantly discussed in Chapters 3, 4, and 8.

In addition, the covenantal fediverse also enjoys contributions from Black technologists, activists, and members (including queer Black folks). But, as several commentators have noted, the fediverse has been racially coded as a predominantly white space, not only resulting in overt racist harassment but also the softer racism of white skepticism toward people of color's concerns.[48] Despite these issues, Black technologists and activists have had a significant impact on the fediverse, as discussed in Chapter 4 and the Conclusion. That Black folks have contributed to this technology is also not surprising—the works of Charlton McIlwain, André Brock, and Catherine Knight-Steele have illustrated the centrality of Blackness to the broader Internet.[49]

## Noncorporate Innovations

A final theme of the book involves social and technical creativity that happens outside of corporate social media—indeed, outside of capitalism. In tech reporting and popular culture, there is a common idea that technological innovation occurs only due to economic incentives, particularly coming from investment capital and monetization. In contrast, the covenantal fediverse has created many innovations. Chapter 1 discusses iterative critical reverse engineering, a way in which alternative social media are developed in relation to corporate social media. As Chapter 3 shows, ActivityPub itself was created without much corporate involvement—a rarity for many technical standards. And fediverse instance admins have developed noncapitalist funding models, refusing to rely on the dominant approach of surveillance capitalism. This theme is most present in Chapter 6. Chapter 7 also explores speculative, experimental innovations happening on the fediverse, particularly in reaction to the climate disaster brought on by carbon emissions—something digital technologies have contributed to. Chapter 8 features resistance to the corporate adoption of ActivityPub. Above all, in the concluding chapter, I discuss the ethical innovation of the fediverse: its emphasis on relationality and the ethics of care. All of these are facets of what "alternative" means.

Mastodon is alternative social media—indeed, it is the most important example of this type of media I have found in my research career. As I have shown elsewhere, alternative social media are built in relation to corporate social media in a process of critical reverse engineering. Mastodon is no different. The next chapter explores how Mastodon was developed in relation to previous social media systems.

# 1

# Critical Reverse Engineering

## How Mastodon Became a Twitter Alternative

As of this writing in mid-2024, most often Mastodon is described in tech reporting as a "Twitter alternative" or "X alternative."[1] This simple statement covers up a great deal of complexity.

First, Mastodon didn't start that way. As Mastodon founder Eugen Rochko told me in an interview, it started out as "just a hobby project" in 2016.[2] It was meant only to be an alternative to an alternative, not a direct challenger to Twitter. Second, Mastodon's emergence as a Twitter alternative was not only the result of Rochko's vision but an ensemble of factors appearing in the early years of the project. These include the decision to license the code as free and open-source software (FOSS), demands for and inclusion of key safety and moderation features, and the impacts of Twitter's declining reputation, particularly among marginalized people. These developments pushed Mastodon into the status of a Twitter alternative by late 2017. Finally, Mastodon is not a singular thing standing in opposition to Twitter. It is at once a project, a codebase, and thousands of small instances. It is also part of a larger network, the fediverse, which includes not only Mastodon and its forks but also a heterogeneous array of other social media applications, all of which have different interpretations of how alternative social media ought to work.

Taken as a whole, Mastodon's early period is a key example of what I call "critical reverse engineering"—the process of critically analyzing other media systems and using the resulting knowledge to produce something better.[3] For Mastodon to be legible as a "Twitter alternative," its designers, admins, moderators, and members had to shape the system in particular ways, both in relation to Twitter as well as other Twitter alternatives. This process is ongoing—so long as there are technologists, activists, and everyday people who say "We can do better," new alternatives will emerge.

*Move Slowly and Build Bridges*. Robert W. Gehl, Oxford University Press. 
DOI: 10.1093/9780197776711.003.0002

## An Alternative to an Alternative

When Rochko started Mastodon in 2016, he had a humble goal: to make an alternative to another already-existing Twitter alternative called GNU social.[4]

Started in the late 2000s, GNU (pronounced gah-New) social was created by Free Software activists.[5] As GNU social cofounder Matt Lee told me in an interview in 2014, "Free software is an ethical choice to protect free expression and self-determination. Any software that is free will carry that to its users, but this does not conflict with the interests of the user."[6] While Twitter allows for particular freedoms—the ability to post, follow, be followed, and like posts across a global network—a major criticism GNU social's Free Software activists leveled at Twitter was that Twitter users could not control the platform itself. This places Twitter users in a vulnerable position. "Twitter can shut down, or go out of business, or they can ban people who use a particular operating system, or they could charge $500 per Tweet," Lee told me. GNU social was designed to address that problem.

GNU social, as I have argued elsewhere, is an example of "critical reverse engineering,"

> a process by which we can take apart and analyze the technical artifacts we confront. . . . This process can be used to critique mainstream social media sites such as Facebook and Twitter. However, this process does not stop at critique; the goal is to use the knowledge gleaned from the analysis of the social media site to make better systems.[7]

I argued that GNU social and other alternative social media "are doing more than decrying what they see as the problems of mainstream social media; they are working to build better systems."[8]

To be legible as a critically reverse engineered version of Twitter, GNU social allows people to do Twitter-like things—write short posts, like, follow other people, and be followed, just as Twitter does. This is coupled with making a better system. The improvements GNU social offered over Twitter are twofold: federation and FOSS.[9] Federation means that there wasn't just one GNU social site, but multiple installations (called "instances"). In fact, there were hundreds of them by 2016.[10] Each of these GNU social instances could communicate with one another via a shared technical protocol.[11] Because GNU social was licensed as FOSS, anyone with the technical ability could download its code, install a distinct instance, and invite friends to join that instance and connect to others. The result was a large network of many small, interconnected social media instances—the fediverse.

FOSS and federated alternative social media are important because they fundamentally challenge our assumptions about how social media needs to work. A basic insight from the field of science and technology studies (STS) is useful here: *it could be otherwise*. Instead of simply accepting problematic systems as the only ones possible, STS argues we should explore what alternative sociotechnical systems can offer. In this way, "STS is a hopeful theory—one that does not simply explain how things have come to be the way they are, but also opens up possibilities for things being ordered in other ways, for the introduction of new and diverse actors, and for rearranging relations of power."[12] Likewise, alternative social media developers open up new possibilities for social media.

GNU social was not alone. Starting in the early 2010s, a host of alternative social media systems were critically reverse engineering corporate social media. In 2010, diaspora* was founded by four New York University students after they heard a speech by a law professor critiquing the governance structure of Facebook. diaspora*—which is still running—was designed to be a noncentralized Facebook alternative that anyone could install on their own server, touting the fact that, when we run our own social media, we're in control of our own data.[13] In 2013, a Brazilian software developer named Miguel Freitas created Twister, a radically noncentralized, peer-to-peer Twitter alternative that ran on bittorrent and blockchain technical protocols. Freitas was inspired by political protests in the Middle East, Brazil, and the United Kingdom but was concerned that protesters were too reliant on centralized systems like Twitter, and that having all activist communications happen in one place enabled powerful government agencies to more easily locate and arrest dissidents.[14]

This wave of alternative social media has inspired yet more projects—and more critical reverse engineering. Eugen Rochko had used federated systems, including GNU social, since high school, and he saw the promise of letting people run their own social media servers. He started to build his own during his studies of computer science at Friedrich-Schiller-Universität Jena.[15] He named his system "Mastodon" after the ancient animal. After graduation, in late 2016, he released a beta version of Mastodon to the public.[16]

Rochko's Mastodon was going to be like GNU social—it would be a FOSS and federated, Twitter-like microblog, allowing anyone with the technical abilities to download it, install it, and connect to other Mastodon instances. Which begs the question: Why make a GNU social alternative? While Rochko liked its federated structure, he found GNU social's user interface to be spartan and unattractive. He put a great deal of energy into the design of the user interface, going so far as to emulate the clean designs of for-profit software, such as TweetDeck. In this sense, Rochko's goal was simply to critically reverse engineer GNU social: he would make a Twitter-like, federated, FOSS system but with a cleaner design.[17]

Having used both GNU social and Mastodon, I do find the latter to be more attractive—GNU social had (and still has) a cluttered aesthetic.[18] I am not the only one. "It's hard to understate how just a small sprinkle of design thinking and caring about the user goes an incredible way in adopting software like this," one blogger noted in 2017.[19] People I interviewed who had been on Mastodon for years often cited its clean design as a key reason they tried it out.

These are the humble roots of Mastodon: a more cleanly designed, critically reverse engineered alternative to GNU social. Rochko critiqued the design of GNU social and made a more attractive version. In doing so, Mastodon inherited many elements from GNU social. But in doing this work, Rochko also set off a process of critically reverse engineering Twitter itself.

## Becoming an Alternative to Twitter

While Rochko intended Mastodon to be an improvement on GNU social, Mastodon's likeness to Twitter prompted people to compare it to that site. Rochko initially discouraged such comparisons. "I'm a realist so I don't think that it will be able to compete with Twitter," Rochko told people on *Hacker News* when he first publicly announced Mastodon in late 2016. "However I would like this project to become the go-to option for people who are already inclined to prefer decentralized/self-hosted solutions, and simply be better than the other software in that space."[20] The "other software in that space" refers to GNU social.

And yet, early on, Mastodon started experiencing bursts of growth—in early 2017, a few months after its launch, it grew by 73%, up to over 40,000 accounts, spread across about a dozen instances.[21] Looking back, Rochko notes this was a sign of what was to come. Mastodon "started gaining traction," Rochko told me in our interview. He started to think "that it had a chance to be bigger than GNU social and actually make an impact."

It wasn't just the design that attracted people. Mastodon does not allow for behavioral advertising. It is not built to engage in what is now known as "surveillance capitalism."[22] This is one of the long-standing frustrations alternative social media developers have with corporate social media: the fact that they gather personal information or monitor people as they socialize, selling personal data to marketers who use it to target ads to people.[23] As a project that critically reverse engineered GNU social, Mastodon inherited an aversion to behavioral advertising from its predecessor. Also, like GNU social, Mastodon does not use algorithms to change the order of posts—instead of being boosted based on metrics like popularity to keep people's attention on the platform, Mastodon presents posts in reverse chronological order.

Even the lack of behavioral advertising doesn't explain why Mastodon grew in the 2016–2017 period. Another inheritance from GNU social is the FOSS model of community development. Mastodon has been fueled by contributions from hundreds of other people. This addresses another concern people have with corporate social media: that they are opaque systems where users have very little control over how corporate social media work. In contrast, because it's free and open-source software, people can see how Mastodon works by looking at its code, and they can contribute changes to that code. While the project started out as Rochko's alone in 2016, within a year, over two dozen other contributors are credited, and the latest version (4.2 as of this writing) has over 1,700 contributors. The number of contributors is evidence that the Mastodon project is capable of and willing to accept code improvements from a large and supportive base.[24]

The community-driven development of these features addresses yet another frustration people have with corporate social media: that corporate executives, not users, are in control of how the services are designed and governed. In contrast, Mastodon allows for far more community control over how the system is shaped—another inheritance from GNU social's reverse engineering of Twitter.[25]

The most important factor, however, was governance. People didn't just contribute code to Mastodon, but ideas on how the Mastodon project should govern itself. As I will describe later, contributors to Mastodon suggested that the project use codes of conduct—standards of community behavior—as well as privacy, safety, and moderation features that were lacking in systems like GNU social as well as Twitter.[26] These include the ability to address posts to specific subsets of people, such as followers only, rather than assuming that everyone wants to post to the entire network. It also includes "content warnings"—essentially, the ability to add a subject line to a post that can either warn people about what's inside (e.g., disturbing content), prevent people seeing movie and TV spoilers, or serve as the setup to a joke.

All of these elements—interface design, its lack of behavioral advertising, the ability of many people to contribute code, and its emphasis on community-centric governance—made Mastodon attractive. In early 2017, Mastodon began to stand out as a stark contrast not just to GNU social but also to corporate social media, particularly Twitter. It was at this point that Mastodon was ready to be seen as a Twitter alternative.

## Twitter and Its Discontents

At that time, there was growing discontent with Twitter. Since 2014, Twitter had been reeling due to #Gamergate, an online campaign against feminists, ostensibly to defend the honor of video gamers. Critics—especially female critics—who pointed out misogyny in both video games and in many gamer practices were targeted, stalked, and harassed with rape and death threats.[27] Then came the United Kingdom's "Brexit" referendum and the US presidential election of 2016. The startling victories of both pro-Brexiters and Donald Trump that year were aided (although certainly not caused) by manipulative social media techniques, including both Russian coordinated Twitter activities as well as domestic right-wing "meme warriors."[28] As media studies scholar Adrienne Massanari argues, there's an overlap between #Gamergate campaigners and the right-wing trolls who support populists like Trump: both support a deeply conservative worldview where women, people of color, and trans folks are meant to be marginalized.[29] All told, there was an ascendant, triumphant right-wing conservatism occurring on Twitter in the mid-2010s—a movement now called the "alt-right."[30]

As *VICE* reported in late 2016, Twitter cofounder Jack Dorsey publicly called for advice from Twitter users on how to make the service better.

> People responding to Dorsey's tweet posed a variety of suggestions, but many asked him to improve Twitter's current status quo for handling harassment. The company recently took baby-steps in this regard; pushing new features like a quality filter and better muting capabilities. But Twitter

> has yet to uniformly enforce broad-sweeping action against hate-speech or the spread of private information.[31]

Essentially, many Twitter users were calling for the platform to ban transphobic, racist, and misogynist hate speech—the speech the alt-right traffics in. They also asked Twitter to ban self-identified Nazis from the site. And they asked for improvements in Twitter's handling of complaints about harassment. "One of the most visible complaints regarding harassment is how infrequently action is taken when a single, regular user files an abuse report," the *VICE* article explains. "Sometimes, Twitter's vetting processes can't even identify blatant hate-speech."[32]

The ascendant alt-right and Twitter's poor moderation of it drove many people to look for alternatives, and Mastodon became a popular option. Mastodon experienced a large wave of incoming Twitter users in early 2017. At that point, Rochko stopped calling it a GNU social alternative and reframed it as something that is better than Twitter.[33] "I brought all my friends to Twitter back in the day," Rochko told *The Verge*. "I kept promoting it to everybody I knew. I really loved the service. But it continuously made decisions that I didn't like. So in the end I decided that maybe Twitter itself is not the way to go forward."[34]

Since tech reporters were largely unfamiliar with GNU social, but were definitely familiar with Twitter, they also started calling Mastodon a Twitter alternative. This happened in a spate of articles that appeared in April 2017. A reporter for *Wired* called Mastodon "a free, open-source, and increasingly popular six-month-old Twitter alternative."[35] A reporter for *Mashable* called it "a potential Twitter-killer."[36] The reporting at this period noted that, unlike Twitter, Mastodon featured strong moderation and safety tools, while still protecting people's ability to express themselves. Perhaps the most quoted headline from that period is from Sarah Jeong's *VICE* article: "Mastodon Is Like Twitter Without the Nazis."[37] "Twitter without the Nazis" became a virtual synonym for Mastodon in its early days. And, at least for the initial Mastodon servers hosted in Rochko's home state, Germany, this was decidedly true, since Germany has restrictions on the sharing of National Socialist content.[38]

These stories accelerated the growth of the Mastodon network. By late 2017, it reached 1 million members spread across over 1,000 individual instances.[39] After the early 2017 coverage, Rochko told an interviewer, "My goals shifted to actually rivaling Twitter for real."[40]

"Rivaling Twitter" entailed more intense critical reverse engineering, this time of Twitter itself. In a 2017 *Medium* post titled "Learning from Twitter's Mistakes," Rochko laid out his critical assessment of Twitter and how Mastodon could be different.[41] He noted that "centralization" or "unexpected algorithmic changes" are problems on Twitter. Those were concerns that both Mastodon and GNU social addressed by being FOSS and federated. But centralization is not the only problem. "Harrassment and tools to deal with it have always been lacking on Twitter's end," he argued. In contrast, Rochko touted Mastodon's safety features, including blocking both individual accounts and entire instances as well as addressing posts only to specific audiences.

By late 2017, Mastodon had shifted from being described by its founder as a GNU social alternative to being reported upon by tech journalists as a Twitter alternative. This was a period of intense critical reverse engineering—Rochko and the rest of Mastodon's developers engaged in critique of both GNU social and Twitter to produce a new alternative social media system. This system, however, wasn't the product of a sole person's vision. It was—and continues to be—a product of struggle among many constituencies: developers, admins, moderators, and members of tens of thousands of fediverse instances.

## Conflicts in Mastodon Development

To show this, let me talk a bit about my own Mastodon experience. I was part of the early 2017 wave of Mastodon members. At that point, I had been studying alternative social media for several years, researching many different systems and publishing articles about them.

I spent most of my early years on Scholar.social, a Mastodon instance that is still online and features an active admin, who goes by the moniker Socrates. Scholar.social has a dedicated moderation team that prohibits racism, transphobia, and misinformation. This was because of Socrates's frustration with content moderation on Twitter.

Socrates self-describes as queer. For him, Twitter "was never a safe space." He talked to me about the false narrative about Twitter: it was a "great public square where everyone had a voice and where you can talk to your friends, you could discuss the main character of the day. . . . And that is absolutely not true. It was only a public square for everyone if everyone doesn't include me, and a bunch of other people like me." In addition, Socrates had no faith in Jack Dorsey, who was the CEO of Twitter at that time. "Jack Dorsey would casually retweet white supremacist stuff," Socrates said, and Twitter itself would not moderate hate speech. He decided in 2017 that Twitter and other corporate social media is not healthy: "I don't like it," he recalled. "I'm leaving the party."

Instead, Socrates, like many other frustrated former Twitter users, switched to Mastodon. And since he has technical ability, he decided to run his own Mastodon instance to create a well-moderated social media space for academics. Even though Scholar.social featured strong moderation, it was well-connected to the rest of the Mastodon network, which was beginning to span much of the globe. Although each had its own distinct moderation standards, the instances Scholar connected to were also well-moderated. For these reasons, I adopted Scholar.social as my home instance in 2017.

I found that my account on Scholar.social allowed me to contribute to a community, provide news about my research, share silly things, make friends and professional connections, and build an international following across the nascent Mastodon federation—all while being in control of my own social media experience in ways that I could not on Facebook or Twitter. While I initially thought Mastodon was just another Twitter alternative, my experience on Scholar convinced me that it might be something more, as well.

As I will show in the pages to come, my research reveals that my happy Mastodon experience was the product of a great deal of hard work and even trauma among the activists, technologists, admins, moderators, and everyday people contributing to Mastodon. Mastodon and the broader fediverse are the product of struggles for ethical social media. For many members of the fediverse, this struggle often included clashes with developers—including Eugen Rochko.

In my interview with Rochko, which was conducted over video chat, he and I bond over a shared interest in playing musical instruments—he sees a guitar hanging on my wall. He tells me he plays guitar, as well, but "I don't put as much time into it as I wish."

That Rochko is hurting for free time is understandable. When he and I talked, it was November 2023—one year since Elon Musk bought Twitter. While the federated Mastodon network had already experienced several waves of new people joining, Musk's purchase of Twitter prompted a massive initial influx and several subsequent smaller ones, depending on any given action from Musk—more on this in the next chapter. Rochko is now CEO of

Mastodon gGmBH, a German-based nonprofit that relies on donations for funding, has several employees, controls the Mastodon intellectual property (the logo and name), directs the development of Mastodon's code, makes apps for iOS and Android, and runs two of the largest Mastodon instances, Mastodon.social and Mastodon.online. He regularly fields requests for interviews from journalists—the period right after Musk bought Twitter was particularly intense. Rochko's time is precious, and he is strained. He told me of a "feeling of burnout and a little bit of depression mixed in there. It's an extremely stressful job" to run Mastodon.

This isn't new. Rochko has reported being stressed due to Mastodon throughout its history. It's not just due to waves of people joining Mastodon due to their discontent with Twitter, although that has happened multiple times since Mastodon began.[42] It's also because Mastodon is simultaneously his project and yet is not under his control.

Since its beginning, Mastodon has been run by Rochko in the "benevolent dictator for life" model, a common approach to FOSS development where a single person makes development decisions.[43] While the project is open to suggestions and contributions from anyone, he makes final decisions about Mastodon software features.[44] In an "Ask Me Anything" session on Reddit in 2022, he wrote, "I do have the final say on the direction of the project. I think that to create a successful product you need a coherent product vision and long-term thinking."[45]

Whether or not someone likes this approach often depends on the vision. Rochko's vision has indeed attracted many people. One of them was Ro, a Black software developer (and major figure in the chapters to come). In 2017, he started his own Mastodon instance, Playvicious.social, because he believed in the vision for self-hosted, noncentralized social media that can be governed by ordinary people. He describes this as "an entirely different digital life that exists outside of what we have been forced to use for so many years," contrasting Mastodon with corporate social media. As he told me in an interview, this form of social media can be "a digital experience that can be so much better with a little bit of effort and a lot of collaboration with love and empathy for the human experience built into its bones." Ro's Playvicious.social instance attracted many people to Mastodon, including people of color.

Likewise, in 2017, queer performer and artist Creatrix Tiara argued that if people wanted "a social network with a heavy queer and trans presence, actually approachable moderators, no ads, considerations for content warnings, and a community-oriented open-source ethos," then Mastodon might be right for them. "Many instances are run by queer and trans people, building safer online spaces for people who have faced major problems online for their identity, including very explicit Codes of Conduct. Queer and trans people have also made significant contributions to Mastodon's code base and development," Tiara writes.[46]

Rochko's vision had attracted many people. But his decisions have also rankled people. Even in the heady early years of Mastodon, Rochko's central role as final arbiter of features grated against many Mastodon contributors. In 2019, *The Daily Dot* reported that many formerly enthusiastic Mastodon contributors and instance members were quitting. The article quoted a Brown queer trans woman who left in "February 2017 after feeling increased alienation from Mastodon's predominantly white userbase" who "made the service 'more and more hostile to the Black and Brown users' that were among Mastodon's initial adopters." Indeed, not long after this *Daily Dot* article, Ro shut down his Mastodon instance, Playvicious.social, due to racist harassment of that instance's moderation team. Ro and many others had been calling for safety features that Rochko did not want to implement—most notably, importable instance blocklists (more on this in a later chapter). That same *Daily Dot* article also described clashes between the queer and trans code contributors and Rochko over who should get credit for the safety and moderation features that were implemented. Later, in 2022, *Fast Company* quoted Tiara, whose enthusiasm for Mastodon had waned: they expressed anger over the harassment of Black and Brown people on Mastodon. Another interviewee cited anti-trans harassment on the network.[47]

At the heart of this is a clash between Rochko and many Mastodon contributors. This has been framed as a tension between open networking—which would build up the nascent Mastodon federated network, a goal of Rochko's—and implementing features that would protect marginalized people from harassment. Indeed, many of my interviewees for this project have a low opinion of Rochko's leadership of the project, criticizing him for making decisions that they believe reduce the safety of marginalized users.

This has been stressful for all involved. In 2018, in a widely quoted post, Rochko erupted at his critics:

> Look y'all here because I built Mastodon the way I wanted and you happened to like it. If you no longer like it that's your own business, don't

give me shit about your failed expectations. There's the door, there's the code, there's the alternatives.[48]

Rochko, like many FOSS developers, was frustrated at the many demands people made upon him. FOSS development is marked by burnout, with developers citing the stresses of addressing often contradictory user concerns, low (or often no) pay, and long hours.[49] Mastodon had shifted once again—it was no longer the result of critique of other systems, such as GNU social and Twitter. It had become itself an object of critical reverse engineering, challenged by many of its early adopters.

## Iterative Critical Reverse Engineering

These eruptions of protest and Rochko's reactions are not an indication that Mastodon is Rochko's fiefdom. Instead, they also show how many people have become deeply invested in Mastodon, arguing for new features, including safety and moderation features. In some cases, their requests aren't accepted. But people keep struggling to get these features implemented.[50] They do not simply accept Rochko's vision for Mastodon. They can take to Mastodon's Github page, request features, and argue for them. They also attempt to cobble them together at the instance level.[51] In some key ways—including, as I will show, codes of conduct and shareable instance blocklists—such suggested features have been adopted into Mastodon. Mastodon itself is being critically reverse engineered by people other than Rochko.

This is a major reason why I use the term "noncentralized" to describe Mastodon. Mastodon is not just the sole vision of Rochko. It is many things: it's a shareable and modifiable code base with hundreds of contributors. It's a nonprofit corporation with a few employees. It's also a network of tens of thousands Mastodon instances, such as my longtime home instance Scholar.social, each with their own membership and community rules, that can connect to each other in a federation. Mastodon may have been started by one university student in Germany, but it is now a result of actions and decisions made by thousands of people around the world.

There is another reason to refer to Mastodon as noncentralized: Mastodon isn't the only federated social media available. In his 2018 eruption, Rochko pointed to "the alternatives." He wasn't telling people to go back to Twitter.

Over the past several years, a host of other ActivityPub-enabled projects have emerged.

Some of these are forks. If people call for a feature in Mastodon and Rochko decides it should not be implemented, they may make alternative versions (called "forks") of Mastodon and include those safety and privacy features.[52] This is possible because Rochko licensed Mastodon as FOSS, allowing others to study and modify its code. Forks of software and projects can often affect the original project, pressuring it to improve.[53]

I talked to Darius Kazemi, a longtime Mastodon code contributor who went on to be the creator of a Mastodon fork called Hometown. Hometown allows members the option to make their posts "local only." "Local-only posting was a hack to get around context collapse," he told me.[54] Instead of having a post travel across the network and thus appear in unanticipated contexts, Kazemi wondered, "What if we just allowed for a certain subset of messages to bounce around in [only within the] server?" This contrasts with Rochko's vision for Mastodon, where posts are meant to move outside of instances and across the network. While this sounds as if Hometown members would be incredibly insular, posting only to their instances and not to the broader fediverse, "it's worked out really well for a number of communities out there who use Hometown. While Hometown is not even top ten for active users, by percentage of active users, we're number five or six," he told me, "Hometown users are highly engaged."

Hometown isn't the only Mastodon fork. There are others, such as Glitch, which includes experimental features. Forks are a major part of FOSS development. Sometimes, features implemented on forks are integrated back into the mainline Mastodon code. Such community experimentation and advocacy have made for a better Mastodon—even if Rochko has sometimes resisted these efforts in the name of a coherent vision and thus avoiding a design-by-committee situation. And, if Rochko does not accept a feature that is seen by many to be important, people can shift to supporting the forks.

In addition, there are a wide range of projects that also run on the ActivityPub protocol, offering fediverse members more options. There are other microblogging systems, such as Pleroma, Akkoma, and Firefish, as well as sites that emphasis longform blogging, images, or music. Each of these systems can innovate new features that improve upon the approach of Mastodon.

Taken as whole, the case of Mastodon, its forks, and the broader fediverse illustrates what might be called "iterative critical reverse engineering." It is "reverse engineering" in the sense that these projects begin by analyzing how other projects work. Mastodon began as an alternative to GNU social, which itself was built as an alternative to Twitter. The forks of Mastodon, as well as other ActivityPub microblogs, are built in part as alternatives to Mastodon. This is "critical reverse engineering" because the intention of these projects' builders is to produce something superior to their counterparts. This is "iterative" because the process is not one-off: Mastodon began as a GNU social alternative, but its members, developers, and tech journalists pushed it to be a Twitter alternative. The process of critical reverse engineering does not stop there, however. Admins and other developers have criticized Rochko's vision for the project. Rochko has had to adapt to the demands of Mastodon members, and in such cases where he does not adapt, there are forks of the Mastodon code or even alternative microblogging systems.

Once a mere hobby project, Mastodon has played a major role in the fediverse since Eugen Rochko announced it in 2016. This period saw steady growth of the network and developments of new features and governance practices, as well as conflicts between Rochko and Mastodon contributors. The result has been a network of thousands of instances comprising hundreds of thousands of people not only participating in Mastodon but also critiquing it. Such critiques can lead to new reverse engineering.

But those early waves of growth would pale in comparison to the stress to come: the influx of millions of people in late 2022, when Elon Musk finalized his purchase of Twitter. I turn to that moment next, focusing on how instance admins dealt with the strain of bringing in millions of new Mastodon members, as well as the difficulties new members faced when attempting to join a noncentralized network.

# 2

# Techlash

## How to Vaporize Elon Musk

One November morning in 2022, Del woke up in her home in South Africa and saw an astonishing number: 2,000.

Overnight, while she was sleeping, 2,000 people had requested to join her social media site, Mastodon.art. Mastodon.art is a fediverse instance intended for creative people who want to share their drawings, paintings, and crafts. It's "a space for artists to converge and support each other creatively," she told me via email. While 2,000 sounds like a small number for a social media site, for a site that had only 19,000 active members, it was a significant influx, far above the daily average of new member requests.

Del is otherwise known as "the Curator," the administrator and operator of Mastodon.art. She had been a member of Mastodon.art since its earliest days in 2016, sharing her art on the site. Her specialty is pen-and-paper style, intricate drawings—mandalas, tangles. A particular favorite of mine is a hypnotic field of skulls and flowers. Within five years of posting her art and participating, she had gained the respect of the community, and when a previous Curator stepped down, Del was tapped to take over. She has been handling the day-to-day operations of Mastodon.art ever since.

The daily life of the Curator of Mastodon.art is normally pretty quiet—helping new members get acclimated to the software, explaining how Mastodon and the fediverse work, providing support to other admins around the world. Del has social anxiety—she gets uncomfortable talking to people in person. But she enjoys online social connection, and she found the work of running a Mastodon server to be deeply fulfilling. She was helping artists from all over the world have a creative home online. Typically, she would have ten to fifteen artists ask to join Mastodon.art in any given day—she would take the time to greet them personally and orient them in the fediverse. Over the years, that day-to-day growth resulted in a social media site populated with 19,000 members.

*Move Slowly and Build Bridges*. Robert W. Gehl, Oxford University Press. © Oxford University Press 2025.
DOI: 10.1093/9780197776711.003.0003

But this period—October 28 through December 2022—was anything but typical. Mastodon.art was swamped with people clamoring to join. All in all, nearly 8,000 new Mastodon.art members joined in that period.

This chapter focuses on this period—a time when a major wave of Twitter users signed up to become members of Mastodon, such as when they joined Del's Mastodon.art. It places that specific moment of Twitter discontent into larger contexts, including the politics of dogmatic free speech absolutism, how corporate social media is sticky—even in the face of controversies—and ways in which Mastodon and fediverse admins and moderators helped would-be new members negotiate a path away from centralized to noncentralized social media. Above all, the chapter further contrasts the power of corporate social media with the power of community-based alternatives.

## The Bird Is Freed

The wave of new Mastodon.art members was prompted by one man. Over 10,000 kilometers from Del's home, on October 28, Elon Musk, the world's richest man, finalized his purchase of Twitter. "The bird is freed," Musk tweeted that day.[1]

Those four words scared a lot of people. Prior to buying Twitter, Musk had declared himself dedicated to free speech, implying that he would relax Twitter moderation rules.[2] In March 2022, he took to his account to poll his followers, asking them whether they felt that Twitter upheld the values of free speech. His followers overwhelmingly answered "no."[3] Many of them decried Twitter's moderation policies, arguing that Twitter suppressed various forms of information, mostly associated with conspiracy theories about the COVID-19 pandemic. This was also in the wake of former US President Donald Trump being removed from Twitter due to the January 6, 2021, insurrection. Musk's followers overwhelmingly argued that unfettered speech on Twitter—implicitly including conspiracy theories and disinformation—is important for democratic deliberation.

The real concern about Musk's free speech absolutism was not over dogmatic equations of speech and democracy. It was over a repeated pattern associated with free speech absolutism on the Internet: it's really a code for unfettered racism, misogyny, transphobia, and harassment. Asked to comment on this point by the United States' National Public Radio, disinformation scholar Nina Jankowicz noted that a Twitter "free speech free-for-all is going to mean less speech for marginalized groups." What happens in free speech absolutist situations, she argued, is that already marginalized people—women, people of color, LGBTQ*, and Indigenous—become overwhelmed by hate speech. Jankowicz herself was the subject of such harassment in 2022 due to a conspiracy theory about her. She received "direct threats to me and my family, my unborn child, people saying that I should be sent to Russia and killed, people saying that I had committed treason."[4] The free speech absolutist position on harassment is that it should be countered with good speech, calling out threats or providing counterarguments (including counterarguments against a conspiracy theory). The problem is that asking marginalized people to face down waves of harassment is only going to further marginalize them.

Anxieties over Musk's potential relaxation of Twitter moderation policies led people to look for another social network to turn to. When Musk tweeted "the bird is freed," in a sense he was freeing many people to ask, "Is Twitter the only way?" Immediately, one might just ask, "Why not use Facebook instead? Or Instagram? Or TikTok?" The odd thing about a dominant, mainstream approach to media is that it tends to become the only imaginable way to communicate. So, if not one corporate social media site, why not another?

However, one major lesson of Musk's takeover is that these corporate sites are at the end of the day controlled by billionaires. Whether it's Musk or Meta's Mark Zuckerberg, why should we trust control of our social interactions to a single person? A cadre of wealthy shareholders (such as the group that owned Twitter prior to Musk's takeover) isn't any better. The Musk takeover was yet another reminder of the need for alternatives to centralized, billionaire-dominated corporate social media.

As the last chapter showed, for as long as there have been corporate social media sites there have been alternatives showing how things could be otherwise. Since the 2010s, systems such as GNU social and Twister offered new ways of doing Twitter-like practices. However, while activists and technologists struggled to make these systems, they never attracted many people to them—even despite controversies in corporate social media.

## We've Been Here Before: The Network Effects of Corporate Social Media

As a scholar advocating for building democratically run alternatives to corporate social media, I have watched growing discontent over corporate social media with varying degrees of hope. Musk's takeover certainly caused a great deal of reflection on the status of Twitter (now X) in society and the utility of people having their sociality and communication flow through a platform now dominated by one billionaire. But I also have learned not to count out corporate social media. Despite outrage after outrage, sites such as Facebook, Instagram, and Twitter are still with us. A tour through the history of social media outrages reveals that discontent does not always mean people will ultimately quit corporate social media.

In 2014 and 2015, #gamergate trended on Twitter, as misogynistic video gamers harassed and threatened progressive, genderqueer, and female journalists and video gamers. Although the campaign was concocted on anonymous forums, Twitter is where the harassment campaign reached full flower, targeting women with rape threats and doxxing, with sock puppet accounts controlled by harassers and trolls. At the time, Twitter's management largely threw up their hands under the belief that unfettered, free flows of information will be a net benefit to everyone. For marginalized Twitter users, however, #gamergate indicated that the social media site did not have their safety in mind.[5]

But people kept using Twitter.

In 2015, there was a tremendous amount of anger over Facebook's emotional contagion study. Facebook researchers partnered with academics at Cornell and the University of California—San Francisco to study how emotion-laden phrases might affect the emotions of Facebook users. In an experimental study—one that no Facebook users consented to—the researchers altered Facebook's algorithms for selected groups of users. Some users saw sad content, and others saw happy things. Then, the researchers observed how the experiment subjects reacted emotionally. When word leaked that Facebook was manipulating the emotions of its users without their consent, Facebook and the researchers were excoriated. Thousands of people declared their desire to leave Facebook for something else.[6]

But people kept using Facebook.

Then, not much later came 2016: the Brexit referendum in the United Kingdom and the US presidential election. As a site that allows for political advertising, Facebook (and its child company Instagram) figured heavily in both events. And at the center of both events was a now-defunct public relations company, Cambridge Analytica, which promised its clients it could use Facebook data to microtarget voters with manipulative messaging.[7] Cambridge Analytica did this work on behalf of the pro-Brexit

Vote Leave campaign as well as the presidential campaigns of Ted Cruz and Donald Trump. Of course, both Vote Leave and Trump won their respective campaigns. After these votes, whistleblowers from Cambridge Analytica stepped forward to reveal that company's manipulative techniques and unethical uses of personal data. One particularly egregious example was the targeting of Black American voters to encourage them to refrain from voting (since Black voters were seen as a reliably Democratic bloc, and thus likely to vote for Trump's opponent, Hilary Clinton). Moreover, the whistleblowers revealed that Cambridge Analytica were able to get their troves of personal data through insecurities in Facebook. Both Cambridge Analytica and Facebook were excoriated for their roles in manipulating democratic deliberation.[8]

But people kept on using Facebook and Instagram. (Cambridge Analytica was dissolved.)[9]

There are two fundamental problems at play here. One is network effects: people use Facebook or Twitter because other people are using them.[10] Corporate social media are useful because they have many people on them—friends, family, coworkers, and artists. Another is that corporate social media are designed to keep our attention. Media studies scholars frame this in different ways, calling attention-grabbing media "sticky" and diagnosing the use of algorithms and dark patterns to keep user attention.[11] Even despite controversy after controversy, many people stuck to Facebook, Instagram, or Twitter—their friends were there, and these networks are designed to keep attention on them.

Will people keep using Twitter, even after Musk's takeover—even after he changed its name to "X"? Is this time different? As of this writing, that remains an open question. One thing distinguishes the 2022 Twitter wave from previous waves of discontent, however: there is a viable alternative available in Mastodon and the fediverse. While there have been alternatives existing alongside corporate social media for over a decade now, the fediverse is a much more mature system, developed by designers who have intensely studied corporate social media for its flaws and opportunities and critically reverse engineered it. Many of the people involved in the fediverse have been working on alternative social media for over a decade. The fediverse also boasts admins who are skilled in moderation and community development. Moreover, as we saw in the last chapter, fediverse admins and moderators had experience with previous waves of former Twitter users switching to the fediverse.

In fact, millions of people did ultimately join Mastodon and the rest of the fediverse in the wake of Musk's takeover, indicating that this wave of discontent could result in long-term shifts in how social media is done, away from centralized, corporate control to noncentralized, democratic control.[12]

## Final Straws and Dead Feeds

Some left quickly, because they took Musk at his word that he would radically relax content moderation practices on Twitter. "When Elon took over Twitter, I became immediately uncomfortable with using it," one person told me. "He wasted no time in letting the public know what his intentions were with the site, and I'm not surprised to discover that the reality of an Elon-owned Twitter is even grimmer than the promise." Indeed, after his finalization of the Twitter purchase in October 2022, one of Musk's first moves was to allow Jordan Peterson and the satirical news site *The Babylon Bee* back—both had been banned for transmisic practices, specifically misgendering trans people. Andrew Tate, the self-proclaimed misogynist influencer, was also reinstated early into Musk's regime. And since the takeover, multiple research teams have found that hate speech has increased on Twitter/X.[13]

Others left because they realized that Musk's post-takeover purging of software engineers—including high-level executives in charge of safety and security—would make the site insecure. "Anyone who has ever witnessed a purge like that before understands the amount of tribal knowledge that has walked out the door," one person told me. "The purge was also done in such a fashion that the people who left were unlikely to offer assistance afterwards. It was cruel and careless." Indeed, in the months since Musk took over, internal Twitter employees as well as whistleblowers are sounding the alarm that the site is far less secure due to Musk's firing of so many employees. Considering how many people post so much personal information to the site—and considering that governments use Twitter/X to communicate with their constituents—this lack of security is dangerous.[14]

Other folks I talked to waited a bit, hoping that Musk's regime wouldn't be as bad as people feared, and that maybe they could in good conscience stay on Twitter/X. But then, a month after completing his purchase of Twitter, Musk invited Donald Trump back to the site. Trump had been removed from Twitter in the wake of the January 6, 2021, insurrection attempt in the United

States.[15] This was the "final straw," one person told me. "At this point it was glaringly obvious what kind of community Musk was chasing and I wanted no part of it. I deactivated my account and moved to Mastodon. Two weeks later I deleted my Twitter account permanently." While Trump has not as of this writing returned to Twitter (he runs his own social media site, Truth Social), what has returned is Trump-style disinformation on the platform.[16]

And still others left because *other* people were leaving. In probably the most painful story I heard, one person—a friend of mine—told me that her Twitter feed simply died the day Musk took over. An avid cyclist, she had been following professional cyclists for years to study her sport. She also followed people fond of sharing cat pictures. This wasn't just about hobbies: posts about bicycles and cats were a lifeline to her when she went through chemotherapy—she would scroll through the pain by looking at bikes and cats. But when Musk took over Twitter, her lifeline was severed. Her Twitter feed died because so many people stopped posting due to Musk's regime.

Overall, Musk's regime has disturbed many former Twitter users over the course of its first year. Twitter/X has fired many employees, thus leaving the site's security and content moderation vulnerable.[17] X now responds to press inquiries with a poop emoji.[18] X offered verified accounts to anyone with $8 to spare (and thus inviting a host of trolls to imitate corporate accounts).[19] It suspended journalists for criticizing Musk.[20] X labeled media outlets such as Canada's CBC and the United States' NPR with the same label as propaganda outlets, such as Russia's RT and Iran's PressTV.[21]

Witnessing all of this, some of the people I interviewed feared a return of some dark periods in that site's history, a throwback to the #gamergate era of racist, transmisic, and misogynist trolls ran rampant, harassing women, Black people, people of color, trans folks, and anyone invested in social justice. Others realized that they had invested much of their online social energy into a site that was vulnerable to simply being taken over by one mercurial billionaire. Still others saw the Musk takeover as the culmination of many other problems—they were already tired of their activities being watched and having ads in their feeds, of cloying corporate brands seeking engagement, of having some of their posts algorithmically favored and others buried, and so when Musk took over, it was as good a reason as any to leave Twitter. "Once Elon took over, it just got scary, so I bailed," one person told me.

These folks looked around for an alternative, and there it was: Mastodon and the fediverse.

They needed help, however. After years of using centralized systems, switching to noncentralized alternative social media proved to be difficult. Recognizing this, experienced fediverse members worked to help those making the jump from Twitter to Mastodon.

## A Little Help from My Fediverse Friends

For those who want to move to the fediverse, there is one major sociotechnical hurdle that stands in the way: Where does one go to sign up in a noncentralized system? A major reason why people stick to corporate social media—despite years of controversies—is that they are easier to join than many of the alternatives, including Mastodon and the fediverse. The problem these would-be fediverse members faced is the fundamental problem people have with federated systems: because they are noncentralized, a would-be member must choose which server to join. By contrast, signing up for Twitter is as easy as going to Twitter.com and filling out some forms. There is no such center in the noncentralized fediverse. Would-be fediverse members must parse a bewildering array of options *before* they can even join and learn how to use a new social media system.

My interviewees, the folks who switched from Twitter to Mastodon in late 2022, spoke about struggling to figure out how to join the fediverse. "It was not obvious or easy at first," one person told me when I asked about the process of switching from Twitter to Mastodon. The interviewees spoke about using Google to search, or simply signing up for an instance because it was the most populated. But most realized relatively quickly that choosing an instance should be about more than its size—it should include things like the instance's focus and moderation.

One key path for many people joining Mastodon is the guide at JoinMastodon.org, which was created by the Mastodon project to promote the service. It's also one of the first results returned on a Google search for "mastodon." JoinMastodon is meant as a gateway into Mastodon, helping people select an instance to join. It lists instances by region and topic. Topics range from general interest to niches like "technology," "LGBTQ+," and "art." Even then, there were choices within choices—twelve LGBTQ+ instances, seven art instances—so which "art" instance should someone join? What distinguishes them? Faced with so many choices, many people no doubt want to go back to what's familiar—the devil they know in the form of Twitter/X.

Would-be Mastodon members needed some guidance, and they got it from people on Twitter itself. During the anxious days after the Musk takeover, Del, the Curator at Mastodon.art, wasn't just looking at requests for accounts on Mastodon.art itself. She also had multiple browser tabs open, including one for her Twitter account. She was watching for people who were considering switching from Twitter to Mastodon.art. "I had a Twitter search for 'Mastodon.art' constantly open in a tab where I'd look for people on Twitter being confused about Mastodon and how to join .art, and I'd help them through the process," she told me. She knew Musk's takeover was driving people away and she wanted to reach them on Twitter to ease their transition to Mastodon. She also knew there was a lot of "confusion about how the different instances work" on the fediverse, and so she wanted to clear things up and help people switch.

Del wasn't the only one promoting Mastodon on Twitter and helping people switch. Twitter was a common way to promote Mastodon to would-be Twitter leavers in late 2022. Multiple interviewees I've spoken to mentioned recruiting people to Mastodon by contacting them on Twitter. When Musk's takeover was reaching its final stages, I also took to Twitter, and, in one of the last things I did with that site before shutting my own account down, I promoted a variety of Mastodon instances (and other fediverse systems, such as Pixelfed, an image-sharing site, and BookWyrm, a book review site) to my followers there. Many other fediverse members were doing the same.

And, as media coverage of the Twitter exodus grew in late 2022, Twitter did something that further promoted Mastodon: Twitter *blocked* the mention of Mastodon on Twitter.[22] I saw this in real time—one day, when I was posting links to various Mastodon instances, an angry red warning box appeared with the text "this link had been identified as potentially harmful."[23] Essentially, Musk's Twitter was using a system intended to help people avoid fraudulent phishing sites as an anticompetition tool, blocking people from mentioning Mastodon and other Twitter alternatives. In the end, however, blocking Mastodon just drew more attention to Mastodon and the fediverse.[24]

All the while, Twitter users continued to discuss what to do about Musk's autocratic regime, and would-be Mastodon members watched their conversations for suggestions about which instance to sign up on. One of my interviewees, an artist named Nina, found Mastodon.art this way, reading deep into Twitter threads on what to do about Musk. "Mastodon came up, and it linked to a few different servers, including .art," she told me. So, Nina may well have been one of the folks Del was watching for on her Twitter tab.

## Even Admins Need Help

As more and more people made the move to Mastodon in late 2022—at one point, hundreds of thousands of people a day—admins like Del started to feel overwhelmed. Del talked to me at length about the work of enrolling so many new members, mornings of waking up and seeing thousands of people asking for access, and days spent teaching new members the cultural practices of Mastodon.art. When Mastodon.art reached 2,000 people requesting access in a day, Del reached her limit: she closed registration for the site to work through the 2,000 applications. "I took the next five days slowly approving those accounts." (Mastodon, unlike a site like Twitter, allows admins to close, open, or limit new registrations depending on what the admin of the server prefers.)

All the while, Del was watching the impact of all the new members on the Mastodon.art server. Using Sidekiq—software designed to monitor and manage computer processes—she could see how her server was processing all the social data flowing through it. Sidekiq offers a graph of how the server was handling things like replies to other posts, boosts, and connections to other servers. Del shared her Sidekiq graphs with me. Normally, Mastodon.art would handle 650,000 such jobs a day. Once Musk took over Twitter in late October, the graph dramatically changed. It reached a peak of 10.8 million jobs—sixteen times the pre-Musk traffic. The load on the server was incredible.

Del wasn't the only one seeing such growth. The explosion of traffic was affecting the entire fediverse of Mastodon and other ActivityPub-speaking servers. Eleven thousand kilometers away from Del, across the Pacific Ocean, Aurynn Shaw was also seeing a massive jump in Sidekiq. Shaw is the admin of CloudIsland.nz, a New Zealand–based Mastodon instance that has been online for three years. Shaw's posts on Mastodon reflect her extensive experience in the tech sector—she's particularly focused on network technologies. She also posts about her experiences as a trans woman in tech. She punctuates these posts with hearty "*ata mārie*!" (Māori for "good morning") posts greeting the Cloud Island community, or sharing some of the art Cloud Island members have made.

During its three years, Cloud Island was easily able to handle network traffic—it was a small server of less than one hundred people. In fact, Cloud Island has what might be seen as a barrier to entry: it's only for paid members, who agree to support the site with a small monthly fee. Even this small site was swamped during the Musk-induced wave of Twitter quitters: the network traffic caused its server to overload several times. Like Del, Shaw was seeing Sidekiq charts explode by a factor of thirteen.

I spoke with Shaw the day after Musk reinstated Donald Trump's account—on November 19, 2022, a month into Musk's ownership of the site. At that point, we had both thought the worst was over, but Trump's reinstatement caused a massive new wave of people moving to Mastodon. As Shaw showed me, Cloud Island saw over a million Sidekiq jobs after the Trump announcement, easily surpassing its previous record from earlier in November. I asked her to talk about what these weeks of Musk waves had been like. "What is time? Time is blurred," she quips. She seemed exhausted. "All of the admins started getting slammed" when Musk took over, she told me, and it was only getting worse. The fediverse was seeing an unprecedented amount of network traffic. "So the load has been . . . kinda *intense* these past few weeks," she says. Fortunately, Shaw has the technical skills to increase the server's capacity, but it still took her two twelve-hour days to complete the task.

It wasn't just network traffic Shaw had to handle. "It's just been a huge difference in terms of the fediverse. There's just so many more people participating." The size of Cloud Island doubled in those days. Again, Cloud Island is small, but the work of bringing in so many new members in such a short time is still intense for a single administrator—that's a lot of "*ata māries*," a lot of orientation to provide.

Shaw and Del—and hundreds of other admins—were all in the same boat. Across the fediverse in late 2022, admins were posting about twelve- to eighteen-hour days being glued to screens, monitoring Sidekiq, handling moderation reports, showing new members the ropes, carefully watching to see what posts and hashtags are trending in the network.

Fortunately, these admins weren't facing those stressful days alone. Del and Shaw are connected, thanks to the fediverse. Their servers had federated long ago, and they got to know each other and other admins, all of whom banded together as they struggled to enroll a wave of new fediverse members. As Del told me, the Twitter influx

> was so big that I started getting help requests from admins of other instances, for moderation advice, help with instance blocks, technical

> help, etc. So I invited a few of them into .art's Discord backrooms, which grew, and now we have I think over 40 admins and moderators in there from different instances, who provide help and support. . . . It's a space for open communication so that we can be on top of potentially problematic situations.

Shaw and Del were able to share their experiences with others, giving and getting advice—and thus mutually handling the stresses of the wave of former Twitter users coming to Mastodon.

The long days weren't thankless, they told me. Once new members joined, both Del's Mastodon.art and Shaw's CloudIsland.nz were welcoming people who expressed relief to be off Twitter. "I've had people tell me they were actually crying when they realised how nice it is here and how horrible places like Twitter and Instagram are for crushing the creative spirit," Del told me. People were appreciative of the work Del and Shaw were doing—creating codes of conduct that actively banned hate speech and building moderation teams capable of enforcing those codes.

My interviewees also praised the content moderation on their fediverse servers. Nina, one of the artists who left Twitter right as Musk took over, found Mastodon.art to be a vibrant and welcoming place. Like many new members, Nina made an #introduction post, talking about her pixel art. "There was a very large response to my post and people were active and talking in all of these introductions," she told me. "It was very high energy, and I was very excited to be able to have a launching off point to interact with lots of cool artists, talking to people that would be future friends, and just a general buzz, compared to the dead silence that is Twitter when you don't have a lot of followers. It was more attention I had gotten in regards to my art for a very long time and made me feel very optimistic about the platform."

Overall, while the Musk-induced wave was a time of struggle—dealing with network traffic, social orientations of new members, and the overall media attention paid to Mastodon and the fediverse—Del, Shaw, and other long-standing Mastodon admins demonstrated their ability to handle the challenge. Their twelve- to eighteen-hour days working on technical and social infrastructures across Mastodon and Twitter helped new members find an instance to join in a sea of noncentralized options. Ultimately, millions of new people joining the fediverse—this remains the biggest gain in membership the fediverse had ever seen.[25]

## "But Can It Replace Twitter?"

However, as soon as January 2023—only two months after Musk's takeover—tech journalists declared Mastodon a failure.

At the height of the wave in November, tech journalists did something I've seen them do many, many times before: they saw a moment of discontent with a major corporate social media platform, they saw people take to a new social media site few had heard of, and they wrote stories about the waves of people leaving Musk's Twitter for Mastodon. (These stories may have prompted Twitter/X to block links to Mastodon for a few days.) I spoke to reporters from NPR, Radio New Zealand, Wired, and the CBC about Mastodon (and, as I stressed in interviews, the rest of the fediverse). Admins, such as Cloud Island's Aurynn Shaw, also gave interviews. It was a heady few weeks.

By far, the most common question we got was, "Can it replace Twitter?" Or, in a bit of a violent metaphor: "Will it kill Twitter?" When it comes to alternative social media, I've seen variations on these questions for over a decade now. For example, back in 2010 when the alternative social networking site diaspora* got media attention, or in 2014 when the social networking site ello was gaining attention, tech journalists would ask alternative social media founders or experts to opine on whether diaspora* or ello would replace or even destroy Facebook. They would then write their stories, declaring the sites "Facebook killers." Then, in a few months, when Facebook was inexplicably still alive, the same journalists would declare sites like diaspora* or ello "failures." I call this cycle of praise and scorn "The Killer Hype Cycle."[26]

The narrative about Mastodon followed the same arc. In January, tech reporters looked and discovered that Twitter/X was still operating despite the existence of Mastodon and the fediverse. Seemingly more alarming was a dip in Mastodon's active user base. Tech journalists wrote stories about Mastodon failing to "kill" Twitter.[27] This is because tech reporting, like any other business reporting, is caught up in capitalist logics of competition for market share. This is the same logic that has us fixated on monthly active users, the global scale of sites like Facebook and Twitter, profit margins and stock prices, and the like.

I would suggest a different framing to tech journalists. Instead of asking, "Can Mastodon replace Twitter?" I would suggest asking this instead: "Is it a viable alternative to Twitter?" This question gets at the fundamental meaning of "alternative" in "alternative social media." Alternative is a not about consumer choices. It's not a matter of choosing Pepsi over Coca-Cola. To suggest that Mastodon could "replace" Twitter falls into the same logic—that we as social media consumers choose among but a few options that are fundamentally the same. I echo what alternative media scholar Clemencia Rodríguez argues: replicating corporate social media metrics does not tell us what is most valuable about alternative social media.[28]

Instead, *alternative* means new ways of thinking about how society should work. What Mastodon and the fediverse are offering are complex, difficult, yet viable answers to the critical questions we've been asking about corporate social media: Does it need to be centralized? Does it need to be under the control of wealthy media moguls? Does it need to monitor our activities and sell our desires and fears back to us? Is a corporation really capable of doing content moderation on a global scale when the network it is moderating comprises hundreds of millions of people drawn from a diverse world of thousands of languages and local practices? Is automated content moderation, or outsourcing content moderation to low-paid, contractual workers, the right way to go? Do we need to obsess over growth?

Above all, the biggest contrasts the 2022 Twitter wave revealed between Mastodon and Twitter/X are the contrasts between individualism versus social groups and centralization versus noncentralization. This chapter began with a common approach to talking about social media: focus on central sites, or better yet, central figures. Here I've focused on Musk, a person mythologized as a self-made billionaire genius (setting aside his coming from a wealthy family) who now owns a major communication channel. He is arguably the single most important person in the history of the fediverse, because his purchase prompted millions to sign up for Mastodon and the fediverse.

Yet the story of Musk's Twitter and Mastodon is actually an inversion: instead of it being a story of the one great man and the big single corporate site, it's a story of heterogeneous networks of people and machines. The fediverse comprises tens of thousands of servers, little instances like Mastodon.art and CloudIsland.nz run by admins like Del and Shaw, and millions of people like Nina and the many people I spoke to who either made the switch from Twitter to Mastodon or were already on the fediverse. Instead of centralizing important people, the fediverse is noncentralized, meaning it is designed to evaporate singular great men—with their singular visions of how social media ought to work—and allow many ordinary people the chance to engage in the struggle to make their own social media. The fediverse is moderated locally, at the instance level, instead of globally, and is done so by humans who are members of the instance communities. It vaporizes Musk and the centralized Twitter in favor of our own, noncentralized social media. This is what it means to offer an alternative to corporate social media.

The ability of Mastodon and the fediverse to absorb the 2022 Twitter wave was a sociotechnical achievement, a potent mixture of social practice with technical infrastructure. Much of this chapter has focused on the social side of sociotechnical: admins guiding people onto the network and sharing tips on handling the traffic, and people working through options of which instance to join.

In the next chapter, I turn to the *technical* side of sociotechnical. I focus on ActivityPub, the technical protocol that enables fediverse servers to communication. The fact that the ActivityPub-powered fediverse was able to absorb millions of new members in a few short months speaks to the technical achievements of the people who created that technical protocol. There's nothing like an influx of millions to stress-test a network.

It turns out that the technical achievement of the Social Web Working Group—the technologists who created ActivityPub—is an important story in its own right. In fact, ActivityPub, a pillar of the noncentralized fediverse, almost never happened. It took the concerted efforts of its authors, a bit of luck, and Mastodon itself to produce this protocol in the face of difficulties and setbacks. With ActivityPub in place, the fediverse has flourished. I turn to ActivityPub next.

# 3
# The Nonstandard Standard
## When ActivityPub Met Mastodon

Amy Guy was doing what many good doctoral students do: procrastinating on their dissertation. Guy, who started studying informatics at the University of Edinburgh in 2012, was looking for any reason not to write a thesis on the Semantic Web, a thesis they had been planning for the past two years.

Like any good procrastinator, Guy was surfing the Web, looking for something else to be doing. Specifically, "I was poking around on the W3C website," they told me. The W3C—the World Wide Web Consortium—is the organization that helps the Web run. It creates technical standards that allow your computer or phone's Web browser (e.g., Chrome or Firefox) to speak to computers across the Internet.

Guy was curious about the groups that make W3C standards. They recalled thinking, "Ooh, what's this working group stuff? What happens if I click these buttons? 'Social Web'—that sounds interesting! What happens if I click 'Join Group'?" After clicking "Join," Guy became a full-fledged member of a W3C standards group, the Social Web Working Group (or SocialWG for short—technical folks like their short names and acronyms).[1]

The Social Web Working Group was chartered in 2014, just in time for it to serve as Guy's ultimate procrastination tool: instead of writing a dissertation on the Semantic Web, Guy got involved in the standards process and helped create the contemporary fediverse.

Years later, looking back, Guy told me, "I had no idea about how any of it worked. I was basically just clicking buttons!" They didn't know that they couldn't just join a standards group without being affiliated with the W3C. It just so happens that the University of Edinburgh was an organizational W3C member. Otherwise, when Guy clicked "Join," they would have just been ignored. The only other path into a W3C working group is to be an invited expert—someone who has expertise the group needs to complete its task. And, as Guy told me, "I certainly wasn't an invited expert!"

*Move Slowly and Build Bridges*. Robert W. Gehl, Oxford University Press. © Oxford University Press 2025.
DOI: 10.1093/9780197776711.003.0004

If there is a standard way to join a standards group, Guy's way certainly wasn't it—it was decidedly nonstandard.

It's fortunate for Mastodon and the fediverse that Guy was able to join the SocialWG, even if a bit naively, because Guy would go on to become a coauthor on one of the fediverse's most important documents: the ActivityPub technical protocol.[2] As other SocialWG members later told me, Guy became an invaluable network builder within the SocialWG, working across different factions in the group, easing tensions, analyzing its decisions, and helping it meet its overall goals.

Drawing on interviews with SocialWG members and analysis of the public SocialWG meeting minutes, this chapter shows how the SocialWG produced ActivityPub. Because ActivityPub enables the contemporary fediverse, it is important to the story of Mastodon. As I show, the creation of ActivityPub was a struggle for better safety tools for queer and trans folks, the struggle of queer and trans software developers to bring their ideas to life even in the face of toxicity and trauma, and the story of what happens when people create social media technologies in the absence of the influence of corporations like Facebook or Twitter.

## Starting at the End

Let's start the story of ActivityPub at the end. ActivityPub is working very well online since it was formalized in early 2018. Millions of people are using it, with more doing so every day. Dozens of alternative social media software projects are relying on it. Because it was created by the W3C, it is an open protocol, meaning anyone can freely build on it.[3]

Like the term "protocol" implies, ActivityPub is a set of technical rules governing the relationship between federation instances. If a given piece of software agrees to the rules of the ActivityPub standard, it can technically join the fediverse. Many small servers, running ActivityPub-compliant software, can connect with one another, making for a massive network of little communities.

Mastodon was the first major implementer of the protocol, implementing a test version in 2017. While Mastodon and the fediverse gained a great deal of attention in 2022 when Elon Musk bought Twitter, these services didn't come from nowhere. Mastodon itself had been around since 2016, years before Musk's purchase. There was already a network of thousands of Mastodon instances being used by millions of people prior to the 2022 Twitter wave. In fact, the network of Mastodon instances had dealt with large waves of new users several times in its history prior to 2022. So, while the 2022 Twitter wave was the biggest one it had seen, it was not the first time.

ActivityPub helped Mastodon to absorb those waves. The protocol allows for Mastodon's noncentralized structure. Rather than a single site being swamped when a wave of new members arrives, members spread around the network of thousands of instances. The ActivityPub protocol allows for the instances of the fediverse to connect to one another.

Mastodon isn't the only software that agrees to the rules of ActivityPub. There are many others, like the image-sharing system Pixelfed, the music-sharing system Funkwhale, or the video-sharing system PeerTube, which run on the protocol. ActivityPub is flexible enough to handle many different types of social media. As I discuss in a later chapter, Meta, the owner of Facebook and Instagram, is implementing ActivityPub on Threads.

Overall, ActivityPub is an impressive work, worthy of inclusion alongside other W3C standards that we use every day on the Web, such as HTML (the language of websites and many mobile apps) and CSS (a design language). But if we look closer at its history, ActivityPub is also a very *nonstandard* standard. This is not just because of things like Amy Guy's stumbling into the Social Web Working Group. There are deeply nonstandard aspects to ActivityPub:

- Corporate social media wasn't involved in the production of ActivityPub. At best, corporations like Facebook or Twitter (as they were known at the time) treated the SocialWG with benign neglect. At worst, they saw the group as utterly irrelevant to the future of social media.
- Instead of settling on one approach to standardizing social media—in other words, a single, standard approach—the group made seven standards. For a typical W3C working group, that is decidedly nonstandard behavior—most groups make one or two.
- ActivityPub almost didn't happen. First, it was considered optional by the SocialWG. An even bigger hurdle came in the face of resistance to ActivityPub from within the SocialWG itself. If it weren't for Mastodon, ActivityPub would not have been completed. Mastodon and ActivityPub connected because of their respective developers' desire for safety for queer and trans people.

- While Mastodon saved ActivityPub from oblivion, it has also effectively erased half of the ActivityPub standard from the Internet. While ActivityPub's creators are grateful to Mastodon for saving their work, Mastodon is also a cause for lament on the part of the ActivityPub authors.

Taken as a whole, it's hard not to see ActivityPub as a nonstandard standard, a technical accident. ActivityPub had no corporate support, almost didn't happen, and—despite its success—people ignore half of it.

I'll explore each of these nonstandard aspects of the ActivityPub standard in turn.

## Corporate Social Media Ignores "Small Beans"

ActivityPub (and all the other standards created by the SocialWG) had zero input from social media corporations.[4] This is extremely nonstandard practice. On the Internet, standards are fundamentally important. Every connection between a computer and the vast network of billions of computers on the Internet is governed by standards. Email is governed by SMTP. Connections to a website use HTTPS. The website itself is likely structured according to HTML standards, and its design—the colors, fonts, and many graphics—will be governed by the CSS standard. These standards structure much of online life, ranging from Web browsing to the connections between organizations sending data around the world.

This means that there are billions of dollars at stake when standards are developed—big corporations often get involved in creating standards. Companies such as IBM, Microsoft, Apple, Google, Meta, Oracle, or Samsung pay membership fees and regularly send employees to standards-setting bodies, such as the Internet Engineering Task Force and the World Wide Web Consortium, in order to shape the standards that govern their industries.[5] As much as corporations like Apple, Microsoft, and Google appear to be competing with one another, in reality they often cooperate when building standards, because the general rule is that standardized technologies are good for business.[6] Corporations that participate in standards setting reduce uncertainty by learning what new standards are in development, enabling them the ability to design products and services that work on those standards before others in the industry.

There is at least one major exception to this rule, though: the ActivityPub standard, the standard that enables federated social media to work. There are many corporations that would have a vested interest in an official World Wide Web Consortium standard that governs the connections between social media servers: Meta (which owns Facebook and Instagram), Twitter, Alphabet (owner of Google and Youtube), Sina Weibo, WeChat, or TikTok.

Given that corporations often get involved in standards production, one might expect some of these firms to be involved with the SocialWG. They were all invited—repeatedly—to join the SocialWG's efforts in creating standards for social networking.[7] None of them showed up. The best the group got in return was a mild expression of interest. As one of the ActivityPub authors, Christine Lemmer-Webber, said, "We were like the only standards group at the W3C that didn't have corporate involvement. None of the big players wanted to do it."[8]

Why? A 2023 story in *The Verge* gives us a clue. When the SocialWG approached corporations like Facebook, Google, or Twitter, "the feedback they got, if they got any, was essentially, we've heard this story before. Federated social was an old idea, and it was never going to work or make any money."[9] Prior to the SocialWG and ActivityPub, there were many federated social media systems. The most notable of these were diaspora*, which was built in 2010 in reaction to the privacy problems of Facebook, and GNU social, a microblogging service created by members of the Free Software Foundation. Still active to this day, both of them are federated systems, allowing anyone to install their software on a server and connect to a larger network.[10] Of these two, diaspora* was arguably the more popular, attracting media attention when it launched—it even received some start-up funding from Facebook's Mark Zuckerberg.[11] However, both diaspora* and GNU social were very small in comparison to Facebook and Twitter, they were challenging to use, and they certainly weren't making money in the same way as Facebook or Twitter.

The lack of profitability is only one part of the answer, however. The other part has to do with a spat between Facebook and Google over open technical protocols. In 2007, Google convened a group of Facebook competitors to create the OpenSocial Foundation. As its name implies, it was a project similar to what the SocialWG would later pursue: creating an open standard by which multiple social media sites could interoperate. OpenSocial's members featured a who's who of late-2000s social media: Google, Yahoo, Friendster, Six Apart, Ning, Bebo, hi5, and MySpace.[12]

Facebook, by all accounts, was caught off guard by OpenSocial.[13] While Facebook wasn't the biggest social media site on the planet in 2007 (that distinction belonged to MySpace), it was growing rapidly and hailed as the hot new site. Within a few years, Facebook would be dominant—Mark Zuckerberg would be *Time* magazine's "person of the year" in 2010—and the other competitors would be left behind.[14] This included OpenSocial. Instead of joining the OpenSocial Foundation, Facebook kept its proprietary application programming interface (API) to itself, and it won the battle with OpenSocial. By 2014 and the start of the Social Web Working Group, OpenSocial would be a shell of its former self, while Facebook dominated social media across much of the globe. Google gave up on OpenSocial, donating its intellectual property to the W3C, and Google's attempts at Facebook-style social media (Orkut, Buzz, Google+) were either defunct or in rapid decline.[15]

By the time of the Social Web Working Group, the big corporate social media players were operating alone, leery of working together. The best that they would do was not prohibit the W3C from pursuing a social media standard. I spoke to Evan Prodromou, who cochaired the SocialWG and was a coauthor of the ActivityPub standard. "Twitter, Google, Facebook, quite a few who are members [of the W3C] could have stopped this process at any time. And there was definitely a sense of they weren't interested in—well, they didn't actually manage to block it, but they *definitely* did not want to participate."

This is not to suggest that the SocialWG didn't try to get these corporations to participate. Noting that Facebook had an excellent social API, Open Graph, the SocialWG, asked Facebook to bring it to the standardization process. "They said 'no,'" says Prodromou. "Twitter's API is also really widely used. We asked them to bring it to the standardization process. They said 'no.'"

All of this makes business sense to Prodromou. Facebook or Twitter's APIs are the "the asset they sell to advertisers. Exposing it out to the rest of the world would be losing that asset, right?" In addition, especially at the time of the SocialWG, Facebook was, as Prodromou notes, a "monopoly . . . they *own* sociality." They were not interested in giving up their monopoly to any open standards group, particularly when they already won the battle against Google's OpenSocial.

The Social Web Working Group was allowed to proceed by Facebook, Google, and Twitter—who weren't going to block it—but without their participation or support. As my interviewees told me, the sense was that the Social Web Working Group was going to fail to have any impact from the start. "None of the big players . . . no one took us seriously," Amy Guy told me. "No one saw [the SocialWG] as a threat. No one expected it to go anywhere. . . . The feeling was, it was small beans, and it wasn't worth their time."

## Standardize Everything

The lack of corporate social media involvement in the SocialWG's work was a mixed blessing. On the one hand, the group was free to conceptualize social media in any way they saw fit without having to address the particular concerns of Facebook or Twitter. As SocialWG member Bengo told me, "I'm glad people from those orgs didn't get involved. It could have made things much more complicated [if they had]."

On the other hand, the lack of corporate involvement also resulted in a lack of consensus among the group members, leading to contention and acrimony. Without the backing and participation of corporate social media—who were the biggest, most successful players in social media—the SocialWG comprised people who had some experience making boutique social platforms, but mostly had theories about how noncentralized social media might be standardized in order to scale up.[16] As Prodromou told me, "Most of [the SocialWG participants] were either open source enthusiasts [or] people coming from an academic or theoretical background." This resulted in the participants having almost too much freedom to reimagine social media, he said. "So a lot of it was kind of witchcraft, right?," said Prodromou. "It's the imaginary world we can have, rather than how do I get more users onto [a corporate] social network."

Within the SocialWG, there were three key factions, each with their own distinct imagination on the best way to standardize social media for the Web. One approach, called Solid, came from the academic research community and was endorsed by none other than Tim Berners-Lee, the creator of the World Wide Web and leader of the W3C. The academic approach envisioned a Web where everything we do online—everything we type, our location in space, audio, video—was annotated in a semantic markup language. In this view, the Web would be a playground for both humans and machines—it would be semantic in that we humans could understand its underlying code, and computers could also read the code and do tasks on our behalf.

Another approach built on existing, smaller-scale federated social media projects. Evan Prodromou, Jessica Tallon, Christine Lemmer-Webber, and Wilkie were key figures here. All had extensive experience building and running federated systems—Prodromou with a social media system called Statusnet and the OStatus protocol (more on this in a bit), and Lemmer-Webber and Tallon with a media-sharing system called Media Goblins. Wilkie had experience making a federated microblog called Rstat.us.

Finally, there was the IndieWeb approach. Led by Tantek Çelik—one of the cochairs of the SocialWG—IndieWeb was (and still is) a growing community of Web developers who placed the homepage at the center of everything. In the IndieWeb world, all of us would have our own websites, with our lives flowing through them: we post our locations, our current status, our goals, our ideas, our lives on our own little part of the Web. These posts would be pushed out to the rest of the Internet (including corporate social media), where other people can check in on us and interact with us through their own websites.

This variety of approaches resulted in a great deal of conflict. Essentially, there were multiple imaginations of a deeply complex phenomena: human sociality. As Lemmer-Webber told me, "I was afraid the group was too fragmented . . . because everyone had their own design." This led to many arguments. As SocialWG participant Wilkie told me, "This is a group of people talking to each other, trying to determine or agree upon how people should talk to each other! I think what's lovely about how contentious it could be—and how passionate people were about this topic—really lends itself to how hard the problem is." Standardizing human sociality is a hard—if not impossible—task.

This contention almost resulted in nothing getting done because factions would dig in, arguing that their approach was the only way forward. As Amy Guy told me,

> For the first—I don't know how long—chunk of time the group was working, the focus was on, OK, let's figure out which one of these approaches we're going to standardize. And all the conversations were focused around that. And as you can imagine, it went round and round in circles. No one could agree, because everyone was invested in their own option. There were a few places where people would not compromise. . . . There were always things that nobody would budge on.

And as Prodromou told me,

> The problem being that there was no downside to failing to come to an agreement. So in my imagined social network world, if you would not agree with it, and not agree to compromise with me, then I don't really care, right? There's no downside to me to keep conversation, to be stubborn, to keep going [with only my preferred approach].

In contrast to the boutique approaches preferred among the three SocialWG camps, corporate social media at least have one common goal in mind: more users, which leads to more personal data to gather and more attention to sell to advertisers. The predominantly academic or theoretical approach taken by SocialWG members refused to accept the corporate approach to social media. This freed the group from the dictates of surveillance capitalism: there were no investors, no executives expecting results, no one demanding infinite growth in monthly active users.

Instead, SocialWG members could explore more philosophical ideas about the meaning of sociality. And this resulted in interminable debates: each of the factions could, if they wish, poke holes in each other's ideas and claim their own approach is best. This led to participants having "trouble keepin' it cool!" as Wilkie told me.

It appeared that the corporate social media representatives to the larger W3C were right—the group was "small beans," not capable of creating anything that could threaten the status quo. But then, as Guy told me, "At some point, we collectively made a decision: just go ahead with everything." That is, instead of deciding which approach to standardizing human online sociality would be best, the SocialWG decided to support everyone's approach. "That basically unblocked us. All of a sudden, people who were at odds with each other were helping each other out. Navigating the W3C process, consulting each other with technical expertise. The whole dynamic changed after that." It was a decidedly nonstandard way to make standards: instead of agreeing on one thing—which would be, of course, a standard—the SocialWG made many standards.

The group would go on to make seven W3C standards.[17] By comparison, I looked at many other W3C standards groups and found that most produced one or two. Seven is a decidedly nonstandard number of standards for a W3C group. As SocialWG member Bengo told me in an interview, "I almost want to say we succeeded too much," that the group made too many noninteroperable standards. But this was the only way forward through the internal conflicts of the group—they had to allow all factions to make their preferred technical protocols.

## Enter the Mastodon: Or, Queer and Trans Technologists in the Mid-2010s

With the conflict between the different working group factions seemingly resolved due to the decision to produce as many standards as possible, there should have been a clear path for what would become ActivityPub, the federated social media standard. There was a faction of people invested in federated social media, including a vocal advocate, Christine Lemmer-Webber, and the tacit agreement was that each faction could get their preferred protocols standardized.

The path wasn't clear, however. Infighting in the SocialWG rears up again as people advocate for and against ActivityPub and the federation protocol. It took intervention from Mastodon to save ActivityPub from the oblivion of being a mere "note"—W3C parlance for something that is not a standard.[18]

The problem was that federation was seen as optional. The SocialWG charter called for the group to produce a "social syntax" (basically, a list of vocabulary terms of online sociality) and a social API (a specification for clients, like mobile phone apps, to connect to social media systems). A social media federation standard was only to be done if there was time, and the Charter called for the group to run for three years, from mid-2014 to mid-2017.[19]

Those who championed federation were often reminded of its optional nature. Going through the meeting minutes, I've observed multiple moments where members of the SocialWG said, "Federation is optional. Federation might not happen"—especially as the group struggled to reconcile its different factions. So, despite the tacit agreement that the SocialWG would allow each faction to produce their protocol, the federation protocol advocates constantly faced pushback.

A particularly clear example happened during a face-to-face meeting in December 2015, when multiple SocialWG members affiliated with the IndieWeb faction, notably SocialWG cochair Tantek Çelik, repeatedly argued that a federation protocol would not happen.[20] Kevin Marks, another

IndieWebber, posted the *Mean Girls* "Stop trying to make fetch happen" meme: "Stop trying to make federation happen," it said. "It's not going to happen."[21]

Two groups fought back against the "federation is not going to happen" view: Christine Lemmer-Webber and her allies within the SocialWG, and queer and trans Mastodon developers outside of the SocialWG. As Lemmer-Webber put it in that December 2015 face-to-face meeting, federation is "the whole reason I'm in the group."[22] She spoke for quite a few people—a courageous act, given that she was saying this as a relatively junior member of the SocialWG and arguing against Çelik, who was one of the group's cochairs and who enjoyed the support of the sizable IndieWeb faction.

Before I say more about Lemmer-Webber's advocacy for ActivityPub, I'll turn to Mastodon's involvement in the federation protocol. Mastodon began in 2016 as an alternative to GNU social, the federated microblogging system. Mastodon was a much more aesthetically and user-focused redesign of GNU social. In comparison to GNU social or other decentralized social media, Mastodon was gaining a reputation as a trans- and queer-friendly alternative social media system.[23]

But it shared something with GNU social, the system it was trying to improve upon. Both Mastodon and GNU social relied on the OStatus protocol. While this meant that GNU social and Mastodon could connect with one another in a federation, this was also a major problem for Mastodon.

The problem was the "O" in OStatus, which stands for "open." Built in 2010, it reflected what one critic calls "Silicon Valley's techno-utopian dream of sharing at all costs"—that is, of total and abiding personal openness.[24] Evan Prodromou was one of its lead developers. As he told me,

> OStatus was based on an idea of sociality that was about public self-expression, which represented a lot of the Web 2.0/blogosphere idea of: this is how the Internet works, everyone can have a voice, and everyone can be their true self in public, and that's the world we want to build. Which was a paradigm which was built by the Tim O'Reillys and the Mark Zuckerbergs and the Evan Prodromous, who have the privilege to live that life, right?

Mastodon's developers—many of whom were queer and trans—did not enjoy the privilege of living the totally open life: they needed privacy and safety, means to protect themselves from harassment. They did not wish to be forced to live in the open.

Initially, Mastodon's developers tried to reengineer OStatus to respect user privacy, giving Mastodon members the ability to set the privacy of each post. But this was done in a crude manner. As Lemmer-Webber told me, Mastodon's implementation of OStatus simply "*marked* things as private and said 'hey other servers, don't peek!'" This proved to be flawed: when these flagged posts started moving across the nascent fediverse, there was nothing to stop others from ignoring the privacy flag and exposing the private information.[25] In fact, the private posts of queer and trans Mastodon members were exposed—often by GNU social instances. Unlike Mastodon, GNU social was developed and run by free speech absolutists who held that everything that can be said, must be said. GNU social was designed to make everything open.

The problem is that GNU social's dogmatic free speech absolutism often attracted right-wing trolls who delighted in violating the privacy of queer and trans folks. Imagine if you were trying to pass a message on a piece of paper to a vulnerable friend across a room, but between you and the friend sat the local bully. The only way to pass your friend the paper note would be to have the bully relay it. It would be naive to hope that folding the note up and writing "Private!" on it would deter the bully from simply reading your note and having more fodder for mocking your vulnerable friend. This is what happened when 2016-era Mastodon members shared private messages. If those messages crossed paths with certain GNU social instances, such as shitposter.club or sealion.club, their contents were shared and mocked.[26] In this way, the openness of OStatus became connected to threats of violence against marginalized Mastodon members.[27]

Mastodon developers learned about ActivityPub, which solved the privacy and flagging problem.[28] ActivityPub's privacy and safety features were there in large part due to the advocacy of Christine Lemmer-Webber, who repeatedly pushed the SocialWG to consider privacy—again, over the protestations of SocialWG cochair Tantek Çelik.[29] Christine's identity as a trans woman impacted her view. Just as "the fediverse has historically been extremely queer," Christine Lemmer-Webber recalled in a podcast episode, ActivityPub was, as well. ActivityPub "was largely developed by queer people—myself included—and including also implementation developers." Four of the five ActivityPub authors self-identified as queer. As for Mastodon, Christine recalls that a large number of its "key contributors, kind of set the tone by pushing for certain features and pushing for kind of certain cultural norms that were to support queer communities that did not feel well-supported by Twitter."[30] So ActivityPub solved many problems for Mastodon—it functioned as a privacy-respecting upgrade over OStatus, a means to break away from harassment coming from some GNU social servers, and a way to further differentiate itself from Twitter.

Mastodon would return the favor to Christine and the rest of the profederation cohort of the SocialWG: Mastodon implemented an experimental, draft version ActivityPub in September 2017.[31] Immediately, Mastodon became one of the largest implementations of any of the SocialWG projects. This is exceptionally important in the world of standards: independent implementation of a draft standard means that the standard is viable, that it's not just the pet project of a W3C standards group member—it could function as a bona fide internet standard.

The impressiveness of Mastodon's implementation was made all the more clear because no corporate social media systems were involved in the SocialWG. In the absence of a Twitter or Facebook adopting a SocialWG standard, Mastodon's implementation of ActivityPub was seen by the group as massive. Mastodon did with a W3C standard what a lot of corporations might do: adopt it while it's still being developed and push out an innovative new idea before the rest of the world catches on. Mastodon's implementation of ActivityPub so impressed the SocialWG that they set up their own Mastodon instance, social.w3c.org—Amy Guy was responsible for this installation.[32]

Most importantly, Mastodon bought Lemmer-Webber and her collaborators (Jessica Tallon, Amy Guy, Evan Prodromou, and Erin Shephard) valuable time to complete ActivityPub. Time was running out: the SocialWG charter specified that the group would dissolve in mid-2017, and ActivityPub wasn't close to being done.

As Guy recalls, the problem was that "the goalposts kept moving. Other specs were getting wrapped up, but there was always a new requirement for ActivityPub that seemed to come out of the blue from one of the chairs of the group. It felt like we were given the runaround." From the outside, looking in at the SocialWG meeting minutes—and I've read them all—it appears that cochair Tantek Çelik was the one moving the goalposts. What his motivation may have been is unclear, but it's remarkable how the other six SocialWG standards got precedence over ActivityPub and how often Çelik discovered new reasons to delay ActivityPub's migration to full W3C standard status.[33]

Fortunately, because Mastodon implemented ActivityPub, even the cochair of the SocialWG could not stop ActivityPub's progress. "The W3C was excited because Mastodon was in the news," Lemmer-Webber told me. "And suddenly we had an extension!" In yet another nonstandard aspect to the ActivityPub story, the SocialWG was granted not one, but two rare deadline extensions—what Lemmer-Webber now refers to as the "Mastodon extension"—to finalize development of ActivityPub. This was largely thanks to Mastodon's use of the draft standard. Instead of ending in mid-2017, the Social Web Working Group wrapped up in early 2018, with ActivityPub being the last of the seven standards to be accepted.[34]

Once seen as optional, the federation protocol was done—and it would go on to be the most popular and successful of the seven standards the SocialWG produced.

## The Ignored Half

The final nonstandard aspect of the ActivityPub standard is that the world pretty much ignores an entire half of it. This has major implications for the future of the fediverse.

As fediverse developer Darius Kazemi notes, the ActivityPub "document actually describes *two separate protocols*. There is a Client to Server (C2S) and Server to Server (S2S)."[35] The server-to-server aspect sets out rules for fediverse instances to "communicate with other servers and propagate information across the social graph."[36] This part of ActivityPub allows for the variety of fediverse social media instances to communicate—a member signs up on an instance and then interacts with that instance to send messages to other people on other instances. It has proven to be very flexible: there are now microblogging servers (e.g., Mastodon, Firefish), video-sharing servers (e.g., PeerTube), and music-sharing servers (e.g., Funkwhale), and each of these can communicate with the rest using ActivityPub's server-to-server rules.

The other half of the ActivityPub specification discusses the client-to-server aspect. This is what is largely ignored, and developers ignore it in large part because of Mastodon. Even though Mastodon adopted ActivityPub and ultimately saved ActivityPub from the oblivion of being a mere W3C note, Mastodon's popularity has also had a negative effect on the protocol.

The client-to-server part of ActivityPub would structure the relationship between, for example, a mobile phone app and ActivityPub-enabled fediverse servers.[37] Many of the SocialWG members really like the client-to-server part. As Amy Guy told me, the client-to-server part was their "favorite bit" of the protocol. They loved the idea that end users could use or create customized clients to access any part of the fediverse. Wilkie, another SocialWG member, also praised the idea, noting that "clients could be catered to particular folks"—this could be extremely useful for folks with a disability, who could include tools like screen readers and other assistive technologies. Above all, Wilkie noted, the client-to-server aspect would not relegate people to using apps that only work with specific services (much like we must today, with a Facebook app and a Whatsapp app and an Instagram app and so on). Bengo, another SocialWG participant, noted that the client-to-server part of the spec could help keep end user data under their control.

None of this is happening, however, because Mastodon's API—its specific version of a client-to-server protocol—is dominant. While Mastodon's size and influence was beneficial to ActivityPub in 2017, saving it from being an afterthought when the SocialWG ended its process, it also has warped the fediverse. Because Mastodon has been the largest ActivityPub implementer, other fediverse services, like Pixelfed, Pleroma, and Peertube, have simply adopted or modified Mastodon's API to build their mobile clients.[38] They want their clients to be interoperable with the dominant fediverse service.

In doing so, they follow the advice of Darius Kazemi. Writing about the client-to-server aspect of ActivityPub, Kazemi says, "Look. If you want to support a universal client ecosystem, go ahead and read this section. But if you just want to write a federated service, skip it and come back to it later."[39] Instead of using the W3C standard, Kazemi argues, it's easier to use the API developed by the first mover in this space: Mastodon.

At this point, no one is coming back to the client-to-server part of the ActivityPub protocol. Another fediverse developer notes, "There is a client to server protocol in the standard, but to the best of my knowledge it is not used anywhere."[40] For all the struggle and last-minute heroics of authors Lemmer-Webber, Guy, Prodromou, Tallon, and Shephard, fully one half of their work has been ignored—even by the fediverse.

The implication of this is that a major part of the fediverse experience—the experience of people who use phone or computer clients to connect to the fediverse—is largely dictated by Mastodon, which was the largest implementer of ActivityPub. As of 2023, however, there is a bigger player involved: Meta, which has launched an ActivityPub-capable microblogging system called Threads. I will discuss Threads's potential to dominate fediverse client APIs in a later chapter.

## After Standardization

Even though Mastodon was the first major project to implement ActivityPub, it's an open protocol, meaning anyone with the technical ability can adopt it. Today, dozens of different projects rely on the ActivityPub protocol, including not only Mastodon microblogging but also PeerTube video, Pixelfed image sharing, and WriteFreely blogging. Tens of thousands of servers run these software packages, and millions of people have accounts on them. All of them can potentially speak to one another, despite the variety of implementations, because of the shared underlying protocol. The variety of types of social media and the large number of servers demonstrate how ActivityPub is a viable technical underpinning of a noncentralized network. As a result, even if Mastodon stopped operating, the ActivityPub-based fediverse would likely continue with the other projects picking up where Mastodon left off.

The biggest indicator of the success of ActivityPub is that it survived the stress test of millions of new people signing up in the wake of Musk's purchase of Twitter. As I discussed in the previous chapter, those new fediverse members caused a massive spike in the network's traffic. While individual servers struggled to keep up, overall, the ActivityPub fediverse was able to handle the load.

These successes have come at a cost. Standards documents don't reveal the stress and toil that its producers went through. There is one last, nonstandard part of the ActivityPub story: the stresses of making a standard may have been higher for its authors, since they did not have the backing of large organizations.

In April 2023, not long after Elon Musk bought Twitter, I asked ActivityPub coauthor Evan Prodromou how he was feeling. I assumed he'd be feeling triumphant. After all, he had been working on federated social media for fifteen years. He was a key part of OStatus, and when his queer and trans colleagues alerted him to that protocol's flaws, he participated in a W3C standards body to produce a superior replacement in the form of ActivityPub. The fediverse's success was very much Prodromou's success. So I prepared myself for a gleeful answer to my question: "How are you feeling about all this success, Evan?"

"Very mixed," he said. He said he was excited about the increasing traffic on the fediverse, "especially the use of Mastodon." But he lamented the loss of jobs happening at Twitter, noting that thousands of people were being laid off by Elon Musk.

And then he added,

> On a personal sense, it brings up a lot of trauma. It's been a long fifteen years. A lot of hard times. It's been very lonely at times. And very difficult personally. It hasn't been an easy place.
>
> There have been colleagues along the way for whom it's been a more difficult path than me. One of the people we talk about in the decentralized community is Ilya Zhitomirskiy, who was one of the founders of diaspora*. We lost him to suicide in the early 2010s. When I think about success, I think about the pressures some of us went through.

Many people struggling for noncentralized, federated social media felt these pressures. Writing to a listserv discussing the future of ActivityPub, Christine Lemmer-Webber noted:

> my experience with the SocialWG especially was that a lack of core agreement on what we were working on really made life incredibly difficult. . . . that took a few years off my lifespan.[41]

Lemmer-Webber doesn't publicly say much more about the toll the SocialWG took on her than this. But she confided in our interview that the stress of building ActivityPub prompted her to consider suicide. I am relieved that she has recovered from that period—the world is a better place with her in it. In the time since the group ended, Lemmer-Webber has largely abandoned the project she championed, ActivityPub. She's joined by other veterans of the SocialWG, including Wilkie, who call for us to move on past ActivityPub.

For their part, Amy Guy was able to procrastinate on their dissertation by stumbling into the SocialWG all those years ago. But now, they tell me that their memory of that time is hazy. They have trouble remembering unless they commiserate with Lemmer-Webber. Guy's recollection is clouded by what "feels like PTSD." "And there are still some things now that will trigger feelings from back then, five years ago," they told me. The happy PhD student gleefully clicking on W3C links now admits that they won't work with certain people who were involved in the SocialWG, and they will never again put all their emotions into a project like they did with ActivityPub.

So the final nonstandard aspect of ActivityPub is that many of the people involved were traumatized by the experience. What kept them going?

Amy Guy's 2017 dissertation, "The Presentation of the Self in Social Media," shows us the ethical commitment the ActivityPub authors had. Now Dr. Amy Guy, they had started their PhD journey looking to write a dissertation about the Semantic Web, a very dry, technical topic. But their procrastination and involvement in the SocialWG ultimately prompted them to switch, focusing instead on the sociotechnical problems of corporate social media and the possibility of building more ethical, noncentralized social media.

Writing about corporate social media surveillance, Guy states:

> I worry about the digital shadow of myself which corporations and governments have access to. I am sure it is thorough and accurate, and that they could use it for all sorts of mischief. I worry about being manipulated without realising, about being tracked, about being backed into a corner with nowhere to hide.[42]

But rather than accept this state of affairs, Guy argues we can build ethical social media: "I think with enough energy and consideration it is possible to build systems which do not do as much harm as the ones we have today."[43] This is not easy:

> The process of creating Web standards, which I engaged with through the World Wide Web Consortium (W3C), involves nitty gritty technical work, understanding obscure specifications, practices, and web lore, and endless pedantic arguments. It is infuriating.[44]

But the payoff, Guy argues, is tremendous:

> I believe (and hope) we are on the verge of an important transition from a world in which our personal data is stored and harnessed by powerful third-parties at great (but often unseen) cost to individuals, to the proliferation of technologies which enable people to be discerning about their choices of communication system.[45]

For all the infuriating and traumatizing difficulty they endured while producing it, Guy, Prodromou, Lemmer-Webber, Tallon, and Shephard's

ActivityPub protocol has helped bring about such a transition. The nonstandard standard has helped bring about an alternative to standard, corporate social media.

The struggle of the authors of ActivityPub was not the only struggle associated with the fediverse. In the next chapter, I'll examine the advocacy of more queer and trans technologists who faced death threats when they called for codes of conduct for technology projects. While they didn't intend codes of conduct to be a key feature of the fediverse, Mastodon's developers created a culture of codes of conduct on the fediverse that has helped give rise to what I call the "covenantal fediverse." The covenantal fediverse is more than just noncentralized: it's a noncentralized network comprising communities that band together through a covenant of shared ethical values. In this way, just as ActivityPub provides a technical protocol enabling federation among many instances, codes of conduct will prove to be a social protocol that helps structure instance governance and instance-to-instance relationships.

# 4
# Codes of Conduct

Shork is in their late twenties, working in software development in Germany. They move data around for corporations—"It's not as fun as it sounds," they wryly tell me. Shork likes video games—particularly retro games on old consoles. When they're not making software or playing games, Shork bikes around their hometown, takes photographs, or hangs out with a local Chaos Computer Club chapter.

But they spend quite a bit of time running Awoo.space, one of the original Mastodon instances. On Awoo, Shork is known by their shark avatar, a cartoon rendering of the Blåhaj toy sold by Ikea. They promise to post "obsessively" about "weird old electronics," and they certainly do—Shork's timeline is full of pictures of old cell phones and proto-smart watches.

Awoo.space was initially set up in 2016 by Eugen Rochko, the original creator of Mastodon, to test out moderation techniques and technologies. Awoo.space took that mandate and went in a unique direction: implementing allowlists. Allowlists on Mastodon are not common. The default behavior of Mastodon is that an instance will federate with another if individual members on those instances follow each other. Federation is typically driven by members, rather than by conscious federating done by admins. An allowlist, by contrast, is where an instance admin only allows their instance to federate with others by prior permission.

That's what Shork spends most of their time working on: whether to allow another instance to federate with Awoo.space. To be allowed to connect to Awoo.space, someone representing that instance must fill out a form, describing the instance and providing some basic information about how the instance is managed.

Why all the caution at Awoo.space? It is because Awoo.space was started and is populated by queer and trans members—including a large population of furries. Furry culture involves a great deal of identity play, shifting from "worksonas" and normalized self-presentation in most settings to "fursonas," self-presentations that identify with animals.[1] "Furries are marginalized," shork tells me. Other Internet cultures tend to shun them. So furries are

*Move Slowly and Build Bridges*. Robert W. Gehl, Oxford University Press. © Oxford University Press 2025.
DOI: 10.1093/9780197776711.003.0005

"protective of their spaces." They wanted a space where they could not feel alone—hence the name Awoo.space. If someone types out "awoo" on the fediverse, emulating the howl of a wolf, others will respond with an "awoo," like a wolf pack.

What is the marker of a fediverse instance that Awoo.space will federate with? As Shork tells me, the most important element is that the applicant's instance has a well-written code of conduct. "The code of conduct is a tool for making it easier for community to organize itself," Shork told me. "Awoo. space requires instances we federate with . . . to have a code of conduct that covers a bunch of basic points." The code doesn't have to be complicated, Shork says. "In the end, it's just a document that states the rules that will make" community on an instance possible.

Codes of conduct make up part of the social and political protocols of the fediverse. While ActivityPub structures how fediverse servers communicate with each other on a technical, networking level, there are a variety of social norms and rules that help structure how people interact on the fediverse. "Protocol" is thus a felicitous term. One denotation is the technical one explored in the previous chapter, the rules that govern computer networking. Another is the rules of personal conduct governing organizational or diplomatic relations. As I will explore here, codes of conduct are rules of etiquette for fediverse server members, and they help shape instance-to-instance relationships in what I call the *covenantal fediverse.*

## Codes of Conduct in Mastodon

Codes of conduct are very important in Mastodon cultural practice. They specify rules that members ought to abide by to participate on the instance. To Shork's point, they don't have to be complicated—they tend to have a few rules. For example, the code of conduct for mstdn.ca, a Mastodon instance intended for Canadians, includes most of the rules typically found across many Mastodon instances:

- No toxic and hateful speech
- No incitement of violence and no promotion of misleading or violent ideologies
- No harassment, dogpiling, or unwanted advances
- No doxxing of other users

- No content illegal in Canada or copyrighted content that you don't own rights to
- Sexually explicit or violent media must be marked appropriately.
- No spamming
- Be yourself.

They also include a bit of elaboration on these rules, a land acknowledgment, and some information about moderators, but all in all, that's it. Many other Mastodon servers include similar information, with some variations based on local regulations or community standards.

Such codes of conduct give an indication of what a new member of an instance might expect when they join. They are declarations of an instance's values and norms, particularly in regard to how members of the instance treat one another. They can vary, but they tend to include language about civility and respect, as well as discussions of consequences for those who abuse others.

They also aid in federation. By default, Mastodon is designed to federate rather easily—if a member of one instance follows a member of another, then the two servers federate, regardless of the existence of codes of conduct. But even so, all instance admins must keep a careful eye on how their server is federating: Are they federating with poorly moderated instances full of accounts that would violate rules, such as "no toxic or hateful speech" or "no spamming"? Codes of conduct become an invaluable guide to deciding whether to keep a connection active.

This can be challenging for new instances. For example, in mid-2023, a Mastodon instance called neuromatch.social included in its About page:

> To other servers: please do not defederate from us yet! We are new, and are attempting a co-operative governance model, so we need some time to figure out how to draft and approve rules as an instance. We have listed some provisional rules for the time being, but the lack of an instance description does not mean we are an unmoderated server: we have mods and admins who are aware of norms within the fediverse, are responsive to reports, fediblock, and other moderation requests, and we intend to do our part in keeping our instance and others that federate with it safe.[2]

Here neuromatch.social recognizes that, without a clear code of conduct in place, other federation instances may decide to defederate, or sever the connection with them. For a new instance, this would radically reduce their ability to reach people across the fediverse. (I'll examine cases in which instances *should* defederate in the next chapter.)

## Aren't Codes of Conduct Just Terms of Service?

At first glance, "code of conduct" might sound like another way of saying "terms of service"—the sort of document Twitter and Facebook use. All social media users are familiar with terms of service agreements—long documents people quickly click through to get to the pleasures of social networking. But codes of conduct are not the same thing.[3] The difference depends on the relationship people have with these systems. Terms of service agreements treat individuals as *users of an isolated service.* People use their service under their rules. In contrast, a code of conduct treats people as *members of an online community, one that can connect to other communities.*[4] A fediverse member is imagined to have some responsibilities toward the community not only in matters of conduct but also in matters of governance—many codes of conduct are cowritten by members of instances.[5] And member conduct on the instance can affect how the rest of the fediverse connects with it.

Simply comparing corporate social media terms of service agreements to typical Mastodon codes of conduct reveals the user/member distinction. Corporate terms of service are full of legalese. While they do cover issues of conduct, they spend a great deal more text on issues like how to resolve disputes with the corporation or how the corporation makes claims to the intellectual property users post. For example, consider X's Terms of Service, which includes these three mellifluous sentences:

> By submitting, posting or displaying Content on or through the Services, you grant us a worldwide, non-exclusive, royalty-free license (with the right to sublicense) to use, copy, reproduce, process, adapt, modify, publish, transmit, display and distribute such Content in any and all media or distribution methods now known or later developed (for clarity, these rights include, for example, curating, transforming, and translating). This license authorizes us to make your Content available to the rest of the world and to let others do the same. You agree that this license includes the right for us to provide, promote, and improve the Services and to make Content submitted to or through the Services available to other companies, organizations or individuals for the syndication, broadcast, distribution, repost,

promotion or publication of such Content on other media and services, subject to our terms and conditions for such Content use.[6]

Congratulations if you made it through that. I should note that it is only a portion of the relevant paragraph: it is only 145 of the over 7,000 words found in X's Terms of Service agreement. This is why people often simply click through them without really reading them: these documents are obtuse and dull.

In contrast, Mastodon codes of conduct tend to be shorter—mstdn.ca's is one of the longer ones, but it's only 1,700 words. One fediverse instance server, friend.camp, cheekily asks would-be members to "Please read the ENTIRE code of conduct. It's like 500 words. I believe in you!!!" Whether it's mstdn.ca's or friend.camp's code of conduct, they are all written to be read, not clicked through. And instead of talking about how user-generated content is to be used, they talk about how members ought to act and how moderators will deal with violations of the code. Rather than being written by lawyers, they are often created by members of the instance. They help members of instances-as-communities govern themselves, and they help Mastodon instances make decisions about federation. They're readable documents, consisting of a few simple rules.

In addition, they don't make claims to the intellectual property posted by fediverse members. This is a major distinction between much of the fediverse and corporate social media: the fediverse is not an advertising platform. It is missing the technologies of surveillance capitalism: intellectual property arrangements that claim a license to people's personal data, massive server farms that store those data, and algorithms that sort those data into demographic and psychographic profiles, and then sell people's attention to marketers.

Overall, corporate social media terms of service pit an individual user—the Facebook, Instagram, YouTube, or X user—against a massive corporation. Those who abide by the rules of the corporation—including its claim to a license to whatever content users post and the use of user data for intense marketing purposes—can keep using the system. Beyond the choice to use the service or not, corporate social media users have no say in how the platform is governed. If a corporate social media user breaks their rules, that person will likely lose access, and their only recourse is either to fill out some form and hope a corporate employee answers, or engage in individual arbitration with the corporation. It's the individual versus a huge corporation, and for the vast majority of people, that's a losing battle.[7]

In contrast, a typical Mastodon code of conduct lays out ground rules for members that guide moderation decisions on a more local, interpersonal or group level. In my time observing the fediverse, I've found that most disputes are resolved through people *talking to each other*, rather than filling in a form or going through forced arbitration. As one admin confided to me, this isn't easy work: there are cases when members are in mental distress, or when a post sits in a gray area vis-à-vis the code of conduct. These moments involve difficult, personal communication. In any case, however, such interaction has the benefit of being personal, a relationship between a moderator and a member. It's not a relationship between an end user and a Web form.

In addition, Mastodon codes of conduct also aid in server-to-server connections. As I will show, the presence (or lack) of a code of conduct helps server admins to make informed choices as their instances federate with one another. In cases of disputes between server admins, codes of conduct provide guidelines for resolution. They thus function as diplomatic documents on the fediverse.[8] This is another contrast with terms of service, which do not specify how Facebook or Twitter/X will interact with other social media services—since, of course, Facebook and Twitter/X don't federate.

Overall, then, codes of conduct are important in Mastodon culture. They aid in helping members of instances know the local rules, and they help instances federate with one another. They are distinct from corporate social media terms of service agreements in that they focus on community building as opposed to corporate/individual relations. They are made by members of the community, for the community, instead of by lawyers for mere users of systems. They help shape federated connections between instances. They comprise the social and political protocols of the fediverse, much as ActivityPub comprises the technical protocol.

But even more importantly, they have their roots in activism against bigotry in tech development. This history has had a massive influence on the development of the fediverse.

## Beyond "Being Excellent"

Codes of conduct are not just important to Mastodon. They are important in the broader tech sector, thanks in large part to the activism of Coraline Ada Ehmke.

Ehmke is a coder as well as an accomplished musician, part of a group called NOSIGNAL, posting videos of her music production to YouTube. In one video, Ehmke wears a Sex Pistols t-shirt and plays a black electric guitar, feeding its signal into a computer. She's working with a collaborator on a cover of a Leonard Cohen song, making it into a dreamy, guitar-laden shoegaze jam that sounds straight off an album from The Cure.

Ehmke has been involved in tech since she was very young, playing with computers in her home in rural Virginia. As she told *WIRED* magazine, technology provided an outlet for her. "Tech completely transformed my life," she said. Because she lived in a small town, "I didn't have a lot of prospects. Tech has brought me great success and given me a lot of opportunity."[9]

But as she got more and more involved in the tech sector, Coraline could increasingly see its harsher side. Ehmke had what she calls a "social justice awakening" when she committed to a gender transition in 2012. "Prior to my transition I was kind of academically aware of around issues of inequity and things like that, especially in the tech field," she told me. "But it wasn't until I began to experience it that it really materialized, and became something like, very real and a lot more relevant to my life." While technology had brought Coraline opportunities, she could clearly see how technology *culture* turned out to be oppressive, particularly to marginalized people.

Coraline wasn't alone. She had many friends, colleagues, and technology activists who helped her think through the tension between technology as a means to expand what's possible versus techno-culture's misogyny, racism, and transmisia. To give back to her friends, and to expand the culture of technology so that more people can have opportunities, Ehmke became one of the world's most important and outspoken tech activists.

Ehmke didn't start small. During and immediately after her transition, she and her allies tackled one of the most sacred rituals in the tech world: the convention, or "con" for short. As anthropologist Gabriella Coleman describes them, technology cons are "a little bit like a summer camp but without the rules, curfews, and annoying counselors."[10] They are places where technologists make and renew friendship. Importantly, they are face-to-face events, supplementing relationships that may have been fostered online.

*Summer camp without rules.* If you think of any 1980s movie set in a summer camp, you can imagine how ribald and silly these cons can get. Much like those 1980s movies—which don't age very well in terms of sexual ethics—technology cons also often had a disturbing side. For years, the *Geek Feminism Wiki*, a resource for feminists in tech-heavy sectors, collected reports of bad behavior at conferences, ranging from keynote speakers making transmisic jokes to soft-porn technology demos to sexual harassment and assault.[11] As a trans woman in tech, Ehmke heard about, witnessed, or experienced these transgressions. And she, and so many others, noted the sheer dearth of women and people of color at free and open-source software (FOSS) conventions.[12]

She and fellow activists campaigned for the adoption of codes of conduct at technology conventions. Rather than have such documents, most technology conventions simply assumed that participants would behave well—that they would just "be excellent" to one another, as a common saying goes. But the growing body of evidence collected by organizations such as the *Geek Feminism Wiki* belied that naive assumption.

Like many activist movements in the 2010s, Ehmke and her allies took to Twitter to pledge not to attend conventions that do not have a code of conduct, using the #CoCPledge hashtag. According to Ehmke, this call led to public discussion. "Many people joined the online discussion—not just advocates and activists, but general members of the tech community. Some shared their personal stories of harassment, even assault, at technology events." This included conference organizers, who wanted to learn more about codes and how they might work. "Several people unfamiliar with the concept of a code of conduct posed thoughtful and sincere questions."[13]

The loudest people in the discussion, however, were far less thoughtful. Some of the response was seemingly polite: "A code of conduct makes us feel good about nice words written on a sheet of paper while no real work gets done," wrote one tech convention organizer.[14] They're too bureaucratic, people complained, and we're here to code. Others were alarmist: install a code of conduct at our convention, they warned, and it will be abused, used to police people's behavior in an Orwellian manner.[15] Others were cryptically racist and misogynist: forcing the tech sector to be diverse would undermine its technical prowess—as if people of color or women can't code. And still others were outright violent: in a disturbing illustration of the very problem she was tackling, Ehmke received transmisic insults and death threats.[16] In fact, she told me that she still gets threats to this day—even when an organization she has no connection to adopts a code of conduct.

This violent resistance to code of conduct activism happens because codes of conduct are a radical challenge to two key pillars of tech work: libertarianism and meritocracy. By insisting upon codes of conduct at conventions, activists were questioning the laissez faire attitudes of the tech community, which saw no reason to restrict behaviors of its members beyond vague statements like "Just be excellent to each other." This libertarian attitude is deeply ingrained in FOSS development. As anthropologist Gabriella Coleman documented in the early 2010s, FOSS communities of practice place a "high premium on self-reliance, individual achievement, and meritocracy."[17]

These ideals are evident in key documents in FOSS culture, most notably in the work of Eric Raymond. In his famous book, *The Cathedral and the Bazaar*, he argues that "in the hacker community . . . , one's work is one's statement. There's a very strict meritocracy (the best craftsmanship wins) and there's a strong ethos that quality should (indeed must) be left to speak for itself."[18] The underlying assumption here is that if everyone is free to write software, then the best software code will bubble up to the top, and that code will be recognized on its merits, rather than being judged by external factors, such as the identity of the author.

In taking on Raymondian meritocracy, Ehmke challenged the common retort to questions about why so few women or people of color attend conventions: they're welcome so long as they can write great software. "But here's the problem," Ehmke wrote in a blog post in 2014. "The majority of today's technologists enjoy elevated privilege in a meritocracy because they have the luxuries of time, money, education, and preferential treatment by the world at large. The assumption of a level playing field is an unfortunate and damaging side effect of a very real lack of awareness."[19] In other words, there are tremendous barriers to even joining a meritocratic field, barriers that prevent contributions from less privileged people. The best work therefore can't even be produced because many people are shut out of the field.[20]

When privileged people are the ones judging merit, they replicate their privilege by anointing their privileged peers as elite. While pro-meritocracy advocates, such as Raymond, suggest that the work speaks for itself, in reality, software code—like so many other human creations—does not lend itself to objective analysis. There are many different ways to implement a solution in code. How do we decide what is best? It's not by judging the code itself; instead, it's by arguments made on the code's behalf by its creators—very often, acrimonious arguments. For a famous example, for many years Linus Torvalds, the creator of Linux, favored an abusive style of argument in technical matters, using statements such as "Kill yourself now" and "Shut the fuck up" in debates over the merit of lines of software.[21] Critics of meritocracy call this "pushyocracy," where the loudest voices often win, even if their code isn't necessarily the best.[22] Adding this to the fact that barriers to entry make the meritocracy an elite club makes for a stagnant culture of self-reinforcement, where already marginalized people don't feel welcome and thus don't contribute software code.

Research confirms this observation. Technologists of color, for example, often face obstacles, including the brute obstacle of bigotry: white coders often question the qualifications of their Black peers. This is in part because Black contributions to science and technology have repeatedly been overlooked. For every movie that reveals Black contributions to science and technology—think of *Hidden Figures*—there are thousands of contributions that are overlooked. Likewise, male coders constantly question the technical skills of women in the field. And in overlooking or belittling the technical abilities and contributions of marginalized people, the meritocratic ideal is revealed to be naive, resulting in the reproduction of technological development as a white, male pursuit, rather than the enriching of technology through diversity.[23]

Most disturbingly, the meritocratic perspective has led to multiple instances where the abusive actions of a skilled coder were often and repeatedly waved away with the excuse "But he does good work." The evidence collected by the *Geek Feminism Wiki* bears this out. The tolerance of repeated racist, misogynist, or transmisic behavior at conferences or in the online settings where coders collaborate leads to women, LGBTQ+, Black, and Indigenous technologists simply staying away from the field.

Ehmke herself would have had multiple excuses to stay away from technology, since she has received death threats. For example, the flagship FOSS project, the Linux kernel, adopted a code of conduct in 2018.[24] Although she had nothing to do with the Linux Project's decision to adopt a code of conduct, it was Ehmke who was harassed. As she told me in an interview, she was doxxed, with her personal information posted online, and she received death threats. She is not intimidated by the threats. She also has a lot of supporters—she's quick to credit all the people who joined the effort. "I owe a lot to other activists in this space that I grew and learned and came up with," she told me.

As a result of the advocacy of Ehmke and her fellow activists, codes of conduct are a regular feature at technology conventions today. In fact, the movement spread beyond the tech sector—academic and business organizations regularly use them. In addition, codes have become a fixture not just in face-to-face settings but are used to govern and moderate online interactions during the development of FOSS. For online projects, Ehmke created "Contributor Covenant," an agreement that all participants in a collective technology development project must abide by. Contributor Covenant includes this pledge:

> We as members, contributors, and leaders pledge to make participation in our community a harassment-free experience for everyone, regardless of age, body size, visible or invisible disability, ethnicity, sex characteristics, gender identity and expression, level of experience, education, socio-economic status, nationality, personal appearance, race, caste, color, religion, or sexual identity and orientation.
>
> We pledge to act and interact in ways that contribute to an open, welcoming, diverse, inclusive, and healthy community.[25]

Contributor Covenant—often appearing with the file name "contribute.md" or "code_of_conduct.md"—is now an essential FOSS governance document, just as important as a software license. Github repositories, where much FOSS development takes place, very often include these documents. This is a massive shift in FOSS cultural and technical practice. Whereas the Raymondian ideal of meritocracy once held for both tech conventions and the online gathering spaces where coders collaborate, today, most conventions and online collaborations are governed by Ehmkeian codes of conduct.

## Enter Mastodon

Marginalized people involved in both technology conferences and online code collaborations have called for—and won—inclusion of codes of conduct in these spaces, but not without a struggle. This took place in the mid-2010s. In the midst of this, a small FOSS project called Mastodon was starting in 2016.

At the time of its start in 2016, Mastodon acquired a reputation as a "Nazi-free" project. Shork, the admin of awoo.space, gives Mastodon founder Eugen Rochko a great deal of credit for declaring early on that Mastodon would not be a home to racists. "That was a thing that attracted a lot of people there," Shork told me. While such a declaration seems straightforward—Who wants racists around?—in the FOSS meritocracy model, the political beliefs or identity of the software coder doesn't matter; the code does. In this view, if a racist could create great software, that racist's work ought to be adopted.

Declaring early on that racists weren't welcome—even racists who were good at coding—drew Mastodon to the attention of queer and trans developers in particular. As I discussed in Chapter 1, these queer, trans, disabled, and furry folks pushed Rochko and Mastodon to be better than Twitter. As Shork told me, "People came to the community with the expectation of—there was obviously a lot of idealism involved—we're doing a new thing, we're trying to do it right."

Since queer and trans developers were already steeping in discussions of codes of conduct in the tech sector, they brought the expectation that Rochko would do more than just say, "No Nazis"—the project should also adopt a code of conduct, explicating the project's values. The software development side of Mastodon adopted Contributor Covenant—the document produced by Coraline Ada Ehmke and allies—in 2017, shortly after the project began.[26]

Importantly, not only did Mastodon adopt a code for the development of the software; it also inaugurated a radical practice: instances themselves, where people would sign up and engage in social media, adopted codes of conduct. The earliest servers, including Mastodon.social and Awoo.space, helped establish this practice in 2017.[27] In doing so, they bucked the normal trend of using a corporate social media–inspired, legalese-filled terms of service agreement, opting instead for social protocols explicated by and for the members of the instances. In addition, the Mastodon project began a Server Covenant in 2019, asking instances to engage in active moderation against hate speech.[28] These actions began and reinforced a cultural trend—to this day, new Mastodon instances are expected to include a code of conduct, and those that don't are viewed with suspicion by more established instances.[29]

Mastodon instance codes of conduct can be created in different ways. Some instance admins simply copy language from the Mastodon Server Covenant or the original instance, Mastodon.social. Many instance admins I spoke to report that they developed their codes in consultation with members of their instances. Still others have opted to create codes and other governance documents in a more formal, collective, cooperative fashion. For example, the Mastodon instance Social.coop, which was founded in 2017, uses software called Loomio as well as a wiki to democratically draft their code of conduct and financial documents.[30]

The adoption of codes of conduct not only differentiated Mastodon from corporate social media sites, such as Twitter, but also from preexisting, FOSS-based, federated social media systems, most notably GNU social. Started in 2010, GNU social had been the flagship FOSS alternative to Twitter. It was developed by members of the Free Software Foundation, the progenitors of FOSS.[31] As such, it was steeped in the very same culture that Coraline Ada Ehmke and allies were struggling against: the libertarian, meritocratic culture where free speech is valued above and beyond civility and ethical consideration of others.

In the previous chapter, we already saw one consequence of this view: the harassment of queer and trans Mastodon members by users of GNU social, who took private posts and openly shared them online, much like a bully might share a personal note passed between friends. Since GNU social had no code of conduct, and since libertarianism means that people are free to do such things, any trolls and racists using GNU social instances tended to face little consequence for such actions. In addition, GNU social as a project resisted Ehmke's call for a code of conduct for many years, only adopting one in 2019.[32] To this day, I have not seen a GNU social instance with a code of conduct—at best, they have a boilerplate terms of service agreement that does not specify behavior.

GNU social isn't the only federated social media system that actively shuns the use of codes of conduct. Pleroma, another Twitter-like microblog which runs on ActivityPub and is thus part of the fediverse—capable of federating with Mastodon instances—also has a culture of eschewing codes of conduct. In many ways, Pleroma has inherited the free-for-all style of GNU social—in fact, the GNU social site shitposter.club (infamous for harassing queer people on Mastodon) is now a Pleroma instance. I have found many Pleroma instances with racist and transphobic content, and only a small number of Pleroma instances with a code of conduct prohibiting those things.[33]

And frankly, that is their right. There is no rule that says a FOSS social media project must specify ethical behavioral practices. In the world of Free Software, "Freedom 0"—the freedom to use software for whatever purpose one wants to—is considered to be an absolute freedom.[34] As Free Software advocates are quick to note, this freedom means that the software can be used for evil as well as good.[35] When this neutrality is coupled with social media practice, the result is extremely light-touch moderation, where only posts that violate laws (e.g., copyright laws, or laws against child exploitation images) are blocked—all else is fair game.

This works in reverse. Because of Freedom 0, a project like Mastodon is also free to adopt a code of conduct. Like the other FOSS projects adopting codes of conduct, Mastodon made a radical move against the dogmatic freedom of FOSS libertarian culture. Moreover, since there is no requirement that *installations* of Mastodon software need a code of conduct, the cultural practice inaugurated by instances like Awoo.social are also a radical move. To attract and protect marginalized people, Awoo.social fused Ehmekian codes of conduct with social media practice. This attracted more and more like-minded people to Mastodon and the fediverse, and they also adopted codes of conduct for their instances to help govern themselves and their connections with other instances.

The social and political protocols contained in codes of conduct have helped bring about a fediverse with a new mode of social media governance: "covenantal federalism."

## Covenantal Federalism

*Contributor Covenant. Federation.* It's a somewhat happy accident we have these concepts. The etymology of *federation* comes from Latin, *foedus*, meaning a diplomatic agreement—specifically, a *covenant*—between two peoples.[36] The selection of these terms by people like Ehmke or computer scientists invokes a long and complex philosophical history: covenantal federalism.

As a colleague and I describe it in a research paper, covenantal federalism involves a covenant created by groups of people who all agree to abide by and be governed under specific ethical principles. Notably, the atomic unit here is not the individual—it's groups of people. In political practice, this might be clans, neighborhoods, or provinces (or, as in the United States, "states"). While each of these groups enjoys a great deal of autonomy, they also consent to be bound together by a covenant.[37]

Covenantal federalism has three main features: *noncentralization*, the *consent of the governed*, and *federal liberty.* Noncentralization is a core organizing principle: it means that the network of groups is designed so that a powerful center cannot emerge. Typically, this is achieved through opposing powerful parts of the system with other powerful parts. The consent of the governed is signaled by the groups accepting the covenant. As for federal liberty, that is the freedom individuals and groups enjoy within the ethical bounds of the covenant.

Political philosophers who advocate for covenantal federalism either look at existing examples (e.g., the federal system in Canada, the Haudenosaunee Confederacy on Turtle Island, or the cantons in Switzerland) or propose ideal versions that could be applied at a global scale.[38] Instances of Mastodon are not states—they don't have armies or coin currency. However, they do provide an example of covenantal federalism in action. Instances (which are most often small communities of people) adopt codes of conduct (which document the ethical principles of the instance) and then federate with other instances with similar codes. This results in an emergent set of widely shared ethical principles across large parts of the fediverse.

The existence of a standard set of ethical principles is clear in the Mastodon part of the fediverse. In 2017, some students and I collected hundreds of codes of conduct from Mastodon instances, and we often saw repeated language used across multiple instances to describe their moderation practices.[39] This has held up over the years: in 2023, a researcher at the University of Warwick used the Mastodon application programming interface to query thousands of instances in order to analyze their codes of conduct, finding that the dominant rules were "no racism, sexism, homophobia, transphobia, xenophobia, or casteism."[40]

Moreover, other fediverse systems are now adopting codes of conduct. Lemmy, a Reddit alternative, now has one for its main project, and many Lemmy instances feature a code of conduct.[41] BookWyrm, a place for discussing books (an alternative to GoodReads), goes one step further. In addition to including a code of conduct on its instances, BookWyrm's code itself draws on an "Ethical Source" license—the anticapitalist license which privileges workers, nonprofits, and educational institutions.[42] When Mastodon, Lemmy, and BookWyrm instances with codes of conduct federate with one another, the covenantal fediverse grows in size and scope.

As for the three features of covenantal federalism, the fediverse demonstrates those, too. The fediverse is noncentralized—it is allergic to powerful centers. An instance's members consent to abide by a code of conduct (or they are free to move to another instance). And even with these ethical guidelines in place, millions of people do in fact enjoy the liberty to socialize with one another—they produce hundreds of millions of posts per month on every conceivable topic.[43]

Awoo.social illustrates this. As an instance that only federates with like-minded instances—instances that themselves have a code of conduct—Awoo.space is the most extreme example of covenantal federalism in Mastodon. However, even as careful as Awoo.space is with federating, it still connects to over 250 other servers—including Mastodon.social, the single largest instance—making it quite well connected with the rest of the fediverse.

In addition, an instance does not need to go as far as using an allowlist. The standard approach—where federation happens as individual members connect to people across instances—works as an example of covenantal federalism, as well. This is because many Mastodon instances keep those federated connections alive when they observe that the instances they are connected to have similar codes of conduct. As one admin told me, "It's very unlikely we will remain federated with an instance without a code of conduct."

In sum, then, codes of conduct emerged from the tech sector in the mid-2010s, thanks to activists like Coraline Ada Ehmke. Much like the technical protocol, ActivityPub, codes of conduct were ready-to-hand social protocols that Mastodon developers adopted in its early days. Spreading across the fediverse, codes of conduct provide protocols for community member behavior as well as protocols for relations between instances. With each instance declaring their ethical values, and instances banding together in part due to shared values, a covenantal fediverse has emerged. This network is an alternative to centralized, corporate social media, which use contractual terms of service agreements to govern users who have little say in how globe-spanning corporate sites function.

## Consent Revoked

Codes of conduct help instances decide to maintain (or allow, in the case of Awoo.space) connections. What about times when an instance wants to sever a connection—or even prevent it from ever happening?

While the fediverse has grown in part through the radical solution of code of conduct–driven covenantal federalism, it also has had moments where instances have been radically isolated. Isolating an instance can be controversial, and until recently, it required a concerted campaign by many admins.

Isolating an instance happens due to the covenantal nature of the fediverse: if an instance blatantly violates the shared ethical norms of the covenant, it is liable to be blocked by the rest of the network. This is a story for the next chapter: the politics of #fediblock and the Black activists who have helped implement instance blocking.

# 5

# Rage and Joy

## Playvicious, #Fediblock, the BadSpace, and the Politics of Defederation

Playvicious.social is one of the most important instances in Mastodon history. For those who have experienced it, it was a joyful place, full of creative and smart people. Its moderation team was deeply respected by much of the fediverse for doing the hard work of fostering a thriving community on the covenantal fediverse.

But Playvicious is no longer online. In 2020, its administrator, Ro, shut it down. His reason: racist harassment.

Playvicious was one of the first—if not the first—Black-run instances on the fediverse. During its three-year run, it featured outspoken Black fediverse members, including Black moderators who tried to keep Playvicious members safe from bigots of the wider fediverse. People of color were rare on the fediverse at that point—arguably, they're still rare. People who consistently and courageously call out racism are also rare. Playvicious members attracted a great deal of hate.

However, while Playvious.social is no more, its influence echoes across the fediverse. These echoes can be seen to this day in two key places: #fediblock, one of the most important hashtags on the fediverse, was cocreated by a Playvicious moderator, and The Bad Space, a blocklist put together by Playvicious admin Ro. Outspoken, courageous Black fediverse members helped create these tools, and the fediverse is better for them.

Taken together, #fediblock and The Bad Space reveal the politics of *defederating* on the fediverse. From one perspective, when two instances defederate with one another—when one or both sever their connection through blocking—it represents a failure of a stated goal of the fediverse: open connection between federation instances and thus the possibility of open communication among all fediverse members. This argument holds that a bigger, more open fediverse will result in more democratic deliberation.

*Move Slowly and Build Bridges*. Robert W. Gehl, Oxford University Press. © Oxford University Press 2025.
DOI: 10.1093/9780197776711.003.0006

But from the perspective of the digital covenant, defederation is entirely justifiable because it protects the integrity of instances-as-communities. There's no reason an instance *must* federate with another if they are both ethically incompatible. If one instance has a code of conduct prohibiting racist harassment and a second one tolerates racist posts, they do not ethically align. The nonracist instance should not be compelled to federate with the instance that tolerates racism. As Socrates, the admin of Scholar.social, told me in an interview, "literally, the only power that an admin has over any other admin is the power to . . . defederate. . . . That option always has to be on the table." From this perspective, the threat of blocking actually keeps the fediverse healthy, because it prevents noxious instances from spreading racist hate speech across the network.

Today, the argument for defederation to protect the covenantal fediverse is strong—so strong, in fact, that Mastodon's developers built the capability to easily block entire swathes of the fediverse into the code. This wasn't always the case. A story of rage and joy—the story of Playvicious, #fediblock, and The Bad Space—tells us how Mastodon came to adopt mass instance blocking. It is another moment in the iterative, critical reverse engineering of the fediverse, where critics of the status quo agitate for improvements in safety and moderation tools.

## Playvicious

As he recounts in his memoirs, Ro started Playvicious.social in a haze of booze and cigarettes.[1] After a long-term relationship ended, he found himself alone in an apartment in Cancun, Mexico, in 2017.

Depressed over his breakup, he was spending a great deal of time on Twitter. Twitter was a source of solace for him, but at the same time he was bothered by it. "[Corporate social media] have some really bad practices when it comes to how they use people's data," he explains in a video blog. "There's a lot of hate, a lot of bigotry being thrown at Black folks, brown folks, LGBTQI, native people." He had already quit Facebook for these reasons, and he was starting to see Twitter become intolerant, as well.[2]

Still, post-breakup Ro was in no mood for figuring out the future of social media. Instead, "I poured a glass of straight vodka, cracked open my laptop, and went back to [Twitter] to see what was happening," he writes in his memoir. "Anything was better than just dwelling and thinking about Sad Things." And while he was there, a friend of his said, "Hey, have you checked out this platform called Mastodon?"[3]

While Twitter was bringing him some company while he spent his days as a self-described zombie in Cancun, it was Mastodon that brought him out of his funk altogether. Ro is an accomplished technologist, delighting in mastering new networking technologies. He dove into Mastodon's documentation and got increasingly excited. It scratched his self-hosting itch. As we discussed in our interviews, he is a big fan of applications we can run ourselves on our own computers instead of relying on big tech cloud services—think of running Nextcloud instead of using Google Docs. He liked the idea that people could install and run their own social media instead of relying on corporate social media sites like Twitter. As he told me, he saw immediately that Mastodon was bringing "the idea of self-hosted social media to the mainstream."

As he writes in his memoir, he was "captivated by creating a self-contained social media platform where I could set the culture. I saw the potential for making an entirely new community without depending on an external service. A place truly for the people and by the people."[4] As he got deeper and deeper into the Mastodon documentation, learning more about how it worked, he put away the vodka in favor of water. He started sleeping more and eating better. And he started a Mastodon instance of his own: Playvicious.social.

The name "Playvicious" (often "PV" for short) came from a collaboration between Ro and one of his friends. For them, Playvicious meant "the energetic, unfiltered, sometimes uncomfortable, but the relentlessly progressive pursuit of understanding ourselves through creativity." It was a space for him and his friends to be creative people. Ro's Mastodon instance, then, was "an experiment [to] find the most genuine expression of self again and build a new community around it."[5]

Playvicious attracted many talented people. Among them was an artist, Marcia X. Marcia knew Ro from Twitter—in fact, Marcia was one of the people who supported Ro through his dark, post-breakup Cancun period. As Marcia told me in an interview, Playvicious was a "godsend" after her time on Instagram and Twitter. It was a place where they could express herself, sharing her photographs, exploring their intersectional Afro-Indigenous identity, rethinking gender and sexuality, and discussing decolonial theory. Because Playvicious was a Black-run instance, it attracted many people of color on the fediverse—although she told me that "there were a lot of white guys on Playvicious." After all, there just "weren't that many people of color on the fediverse at that time, and there still really aren't that many."

In keeping with the name "Playvicious," Marcia brought a great deal of creative energy to the instance. They started hashtags, like #FineFemmeFriday to highlight self-identified women, #FridayFeelsRadio to share music for the weekend, and #Fedikitchen for the foodies of the fediverse. In our conversation, Marcia downplayed these contributions in comparison to coding—"I don't do tech. I don't do code," she told me. However, her contributions to the culture of the fediverse are just as valuable as its underlying software: these hashtags—and others she created—are still in use.

When he talks about Marcia's efforts to make Playvicious and the rest of the fediverse a better place, Ro absolutely gushes. "In terms of how she interacts with culture, how she interacts with people, how she expresses ideas, how she has the ability to take this piece of information, this culture, this practice, and meld it into a tangible idea, and able to express it. . . . I know a lot of smart people. But her? She's somewhere else," he told me. "She is absolutely brilliant." Ro found himself relying on Marcia more and more for advice on how to run the instance. "She was really the one I confided in and leaned on in terms of setting the tone for the culture [of Playvicious]." Soon, Marcia X was promoted to moderator of Playvicious.

Overall, 2017 was a heady time for Playvicious. Marcia was leading conversations into complex cultural topics. Ro remained positive about Mastodon and the fediverse. "I love the fediverse," he wrote on Patreon that year. "I've been a huge fan since I joined everybody on Mastodon with my own instance."[6]

## Enter the "Harassing Dickheads"

But the fediverse wasn't always loving Playvicious.social back. In 2018, Ro wrote about "harassing dickheads" he had been staving off.[7] This was an early harbinger of things to come.

Before I go further, I need to say a bit more about Ro. He's not only willing to tackle any digital technology and master it, he's also dedicated to working on his social interactions to become a better, more thoughtful and kind human being. In our interviews, he talked to me about the importance of seeing a therapist and exercising. He often gives advice at the end of his videos: "Drink more water, get some sleep, see the doctor, take care of yourselves." He is a kind and joyful person.

Unless he's interacting with bigots. He has no tolerance for them. "I know how to deal with bigots on the fedi," he told me. He enjoys "fucking with them." He almost sounds gleeful as he talks to me about this. Indeed, in my time on the fediverse, I've observed Ro argue with racists many times.

Since Playvicious was an unabashedly Black Mastodon instance, it attracted plenty of bigots for Ro to argue with. But while Ro enjoyed battling against bigoted trolls, he also recognized that Playvicious moderators—people like Marcia—did not want to. Marcia wanted to talk about philosophy and art, not stave off racists. Moderating against racist harassment was taking a toll on Marcia and the rest of the Playvicious moderation team.

Still, the fediverse seemed to provide more positives than negatives for Ro, Marcia, and Playvicious. It just needed better moderation tools. Ro started thinking of ways to improve moderation technologies in Mastodon, looking for ways to make Playvicious safer from trolls. One key idea he had (and he was not the only one) was for Mastodon admins to be able to export, share, and import instance blocklists. That way, they could block racist instances en masse.

At that time, if an admin wanted to block another instance, they had to enter the offending instance's URL manually into their moderation interface. This is not a high burden if there's one bad instance to block. But Playvicious was discovering one of the major downsides to easily self-hosted social media: the ease with which someone can set up an instance, which can result in many problematic instances to deal with at once. Multiple admins told me about "predatory instances" on the fediverse that seek out new instances to harass. As Scholar.social's Socrates told me, the low cost to entry for federated social media server software leads to such instances popping up constantly: "You buy a cheap, cheap domain name, you spin up a cheap, cheap server, put Pleroma on it, boom, 20 minutes later, you're calling people the n-word."

Particularly during waves of fediverse popularity—such as when a major news outlet writes a story about Mastodon or the fediverse—"free speech extremist" instances pop up like mushrooms. Operating under the dogmatic logic that anything that can be legally said ought to be heard, free speech extremist instance users delight in dropping into people's threads and insulting them. As a fediverse admin by the name of Oliphant told me, the very first response he received to his #introduction post was "No one cares, tranny n----- faggot." There are hosts of trolls watching for introduction posts, looking to harass people. When such waves occur, adding instances one by one to one's own server blocklist is inadequate.

Mastodon members started calling for the ability to import instance blocklists as early as November 2016.[8] Variations on this idea appeared in Gitub Issues and Mastodon's Discourse discussion board again and again. It ended up being debated for almost *seven years*: instance blocklist importing was only implemented in February 2023, in version 4.1.0 of Mastodon.

What accounts for this delay in implementation?

There were debates about the wisdom of implementing such a feature. My participant observation and interviews reveal four main arguments people have put forward against the ability to import blocklists:

- *The dogmatic free speech argument.* This argument holds that blocking people preemptively violates the principle of free speech. Rather than block a server that houses someone who holds bad views, instances should meet their bad speech with good speech in open debate.
- *The open federation argument.* This is related to the free speech argument. The open federation argument says that blocklist importing would break the federation, because it would sever too many connections between fediverse instances and isolate people from one another. Since the fediverse is meant to be a vast, open network, breaking connections should not be done at the instance level—it should solely be the choice of individual users on a case-by-case basis.
- *The grudgelist argument.* This argument holds that the most popular people will create the most used blocklists, and no matter how virtuous that popular person is, they will inevitably put instances on the list that don't deserve to be there due to grudges or interpersonal conflicts. Once an instance is on a list, the fear is that it can never be removed.
- *The responsibility argument.* This is related to the grudgelist argument. This argument holds that responsibility for blocking falls on individual users and admins, and they should do their due diligence before blocking anyone. By importing a list, they are offloading this responsibility to a third party and are thus being irresponsible by taking the easy way out.

There were (and still are) many people opposed to implementing importable blocklists, but the most notable resistance to blocklist importing came from none other than the founder of Mastodon himself, Eugen Rochko. While he was not in favor of the dogmatic free speech argument (after all, as we saw in the previous chapter, he quickly agreed with the call for codes of conduct that push back against the dogmatic free speech position), he often invoked the last three arguments. "I don't think shared server blocklists are a good idea," he wrote on Mastodon in 2018. "It's better for every admin to examine the accuracy of the decision." He went on: "otherwise something like the Randi Harper situation can happen again but on a much larger scale."[9]

In that single post alone, Rochko implicitly or explicitly invoked the last three arguments against shared blocklists. He invoked responsibility by noting admins must do due diligence for every instance block. He invoked the grudgelist argument by citing the case of Randi Harper, who contributed to Twitter blocklists of #GamerGate trolls but also placed many trans artists on a popular blocklist, leading to them being lumped in with alt-right GamerGaters and effectively ending their careers.[10] And he invoked the open federation argument by noting that shared blocklists could cause Harper-like grudge-blocking on a massive scale, breaking up the fediverse.

Rochko's resistance to importable instance blocklists was the ultimate argument against the feature, because what Rochko says goes. Throughout the life of the Mastodon project, Rochko has maintained a "Benevolent Dictator for Life" organizational structure.[11] Ultimately, then, what accounts for the seven-year gap between a call for the ability to import instance blocklists and the implementation of the feature is that Eugen Rochko did not want that feature in Mastodon.

This was quite disappointing to Ro. In light of the toll that racist trolls were taking on moderators like Marcia, he could not understand why safety features, such as importable instances blocklists, were not being implemented. From Ro's perspective, Rochko's vision has actively prevented safety features from being a part of Mastodon. "My biggest disappointment with Mastodon is the development ethos itself—it kind of refuses to adapt," he told me. "It's the singular vision of one person. And that's fine, if it's a project for you and your friends." But when it's a project meant for many different communities, the "lack of consideration puts a lot of people in danger," he says—marginalized people especially.

## #Fediblock

While calls for an instance blocklist importing feature were denied by Rochko for seven years, bad instances full of racist and transmisic trolls kept appearing on the network. People could not wait for the implementation of shared blocklists to figure out how to coordinate efforts to block bad instances.

In the absence of shared blocklists during that seven-year period, Mastodon and fediverse admins have relied on organizing via hashtags. The most important of these is #fediblock. One of its creators is Marcia X of Playvicious.social. "I started [#fediblock] when I was still on PV," she told me. She wanted to make a public forum for discussions about bad instances and abusive accounts, but at the time, most of these discussions happened in private. "Because so much of what we had done was behind the scenes between DMs and group chats with ourselves, I kinda started to get frustrated." Marcia felt that open, transparent discussions about bad instances would be far more effective in increasing the safety of the fediverse.

She built the hashtag in collaborating with a fellow Mastodon friend, Ginger. I asked Ginger about #fediblock. She said,

> We did not (and still do not) have effective moderation tools, not even the note section was available for us to remember why we blocked someone/an instance. We were using Excel sheets to try to keep track of everything. I think it was Marcia's idea to use a hashtag specifically, so it could be seen/accessed by anyone who needed it, when they needed it.

#Fediblock became—and still is—an essential tool on the fediverse. As Marcia told me, the hashtag "took on a life of its own, and people really took to fediblock and it's almost a really important part of most instances."[12] For instance admins, one key piece of advice is to follow that hashtag, making sure that racist or abusive instances are blocked immediately. In fact, when I become the admin of a Mastodon instance myself, one of my first acts was to follow the #fediblock hashtag, paying careful attention to the discussion there and blocking based on that discussion. This choice has paid off, I believe, with a safer Mastodon instance.

However, like many hashtag movements, its original authors would often get overlooked. As #fediblock grew in importance, multiple projects took the hashtag to create public #fediblock lists (again, until only recently, anyone wanting to draw on those lists would have to do so by adding them one by one). One project, FediBlock.org, did not credit Marcia or Ginger for their work. But they did build on their work by creating an interface for people to suggest instances that should be blocked and publishing the resulting list.[13]

When the original authors of #fediblock, Marcia and Ginger, *were* publicly noticed, the attention was often negative. As an Afro-Indigenous woman, Marcia was a favorite target. Marcia told me about being stalked, with people using multiple accounts, screenshotting her posts, and sharing their personal information. When this happens, "I have no protection," she told me. "I'm just a guy using a computer!"[14]

While Marcia and Ginger were either being erased in fediverse history or attracting unwanted attention, their #fediblock hashtag has lived on—and hashtag-based coordination has some major triumphs to its name. The power of hashtag-based coordination against bad instances could be seen in the isolation of one of the most notorious instances in Mastodon history: Gab.com.

## Blocking Gab

Founded in 2016, Gab.com describes itself as "The Free Speech Social Network and Pioneer of the Parallel Economy." Its founder, Andrew Torba, positions his site as a defender of free speech and an antidote to cancel culture—those who have been removed from another media platform should come to Gab, he says.[15] Judging from the biographies of the accounts I found when I studied the site in depth in 2022, this call to the deplatformed seems to have resonated. Many of them claim to have been banned by Twitter, Twitch, or other platforms.[16]

But in my analysis of Torba's writings about Gab as well as the content produced by its users, it is clear that Gab is not meant for just anyone. Gab is a white Christian nationalist social media site. Kristin Kobes Du Mez, author of the book *Jesus and John Wayne*, describes white Christian nationalism as a "God-and-country faith" that is "inextricably linked to a staunch commitment to patriarchal authority, gender difference, and Christian nationalism, and all of these are intertwined with white racial identity."[17] White Christian nationalists believe cisgender, white men must be the manly protectors of both home and nation. Such men must dominate women, while at the same time bending the knee to a Jesus who is a "conquering warrior, a man's man who takes no prisoners and wages holy war."[18]

Torba's writings make this clear. "We need Christian men who embrace their God-given masculine energy to conquer and lead," he writes.[19] Non-Christian women are "whores"—Christian women obey their master men.[20]

Trans people should not exist, Torba argues. Big Tech, he argues, is run by a conspiracy of international Jewry. If all of that wasn't enough, Torba argues cities are full of filthy people of color who are trying to replace white people, and that Donald Trump won the 2020 election.[21] Torba's Gab has cultivated a userbase that often shares these values. In late 2022, I studied the top accounts on Gab, finding among them politicians such as Matt Gaetz, Margorie Taylor Greene, and Lauren Boebert, people selling QAnon conspiracy theory books, evangelical pastors who tout both Jesus and masculinity, doctors who claim trans people are sick, right-wing news aggregators, Great Replacement theorists, and self-identified Nazis. A Pew Research study in 2023 echoes what I found.[22]

What does Gab have to do with the fediverse? After failing to build his own social media software in the years 2016 to 2019, Torba realized he didn't need to start from scratch: he could just use Mastodon. Since Mastodon is free and open-source software (FOSS), anyone can take its code and run it. Torba found Mastodon to be extremely attractive because of its use of ActivityPub, the technical protocol that allows the fediverse to run. In theory, Gab could join the fediverse and quickly connect to millions of people across thousands of other instances. On July 4, 2019—Independence Day in the United States—Torba opened up the new Gab, declaring that it would bring about a "free speech federation" of other white supremacist sites. "Gab is now unstoppable and can never again be taken down as a whole ever again," Torba claimed in an interview with *VICE*.[23]

The battle cry of Gab's free speech dogmatism is, essentially, you need to hear everything, including hate speech. "A speech without an audience is just a monologue delivered to yourself," Torba argues. And if you don't have an audience, you don't matter. "Jesus has an audience, so the Jews killed him. The Jews did not care about His freedom to speak, they cared about the amount of people He was reaching with that speech," he writes. "The same is true today. The Jews in positions of power do not care that you have the freedom of speech in this country."[24]

As we saw in the previous chapter, open debate has been traditionally amenable in FOSS culture. Similarly, this vision of open discussion resonates with a number of fediverse members who held the dogmatic vision that bad speech should be met with good speech in open debate. Torba's adoption of Mastodon's software led him to believe that the sorts of racist, alt-right conspiracy theorizing that Gab housed would be "unstoppable" and would quickly spread across the fediverse.

However, people like Marcia, Ginger, Ro, and other like-minded fediverse admins and moderators have another perspective, based on harsh experience: they recognize Gab's free speech absolutism as an example of what philosopher Herbert Marcuse calls "repressive tolerance." Repressive tolerance is inherently conservative: it is a passive acceptance of the status quo. By placing oppressive or violent ideas on equal footing with marginalized ones—such as in the case when racists enter debates with antiracists—the radical ideas of the marginalized are drowned out in a deluge of the status quo position. As Marcuse argues, repressive tolerance is "the passive toleration of entrenched and established attitudes and ideas even if their damaging effect on man and nature is evident." It is "the active, official tolerance granted to the Right as well as to the Left, to movements of aggression as well as to movements of peace, to the party of hate as well as to that of humanity."[25]

Repressive tolerance is the kind of tolerance that makes the dogmatic free speech position dangerous. Placing dominant ideas (such as white supremacy, patriarchy, evangelical Christianity, and heteronormativity) on the same footing as marginalized ideas ensures the eradication of the latter by the former.[26] Moreover, asking people of color or trans folks to defend their very existence in online debates only further entrenches white supremacy and conservative sexuality, because that burden unevenly saps their energy while non-Black and cisgender people can blithely continue on with their daily lives. The dogmatic free speech position would tolerate Gab members' supremacist ideas in the name of open debate, even if those ideas are already dominant in society as a whole. And this tolerance would result in the repression of people who challenge supremacism.

Fediverse admins who recognized the threat of Gab's entry into the fediverse used hashtags to coordinate a blocking campaign against Gab. As Derek Caelin, who wrote a history of this moment, writes, "The migration of Gab marked a major test for the fediverse. Because no one authority controlled the policies of the network, it would be impossible to bar Gab with a single sweeping action, the way centralized social networks such as Facebook or Twitter might. Instead, it would take a mass movement of administrators, moderators, and technology providers to block, isolate, and moderate the new server."[27] Like-minded instances—the covenantal fediverse who shared similar codes of conduct—banded together via hashtags, coordinating a campaign to isolate Gab.

They were effective. Gab was blocked from the vast majority of the fediverse. In fact, it was so well blocked that Andrew Torba indulged in what could be characterized as a childish rage-quit. In mid-2020—not even a year after pronouncing his desire to create an "unstoppable" free speech federation of white Christian nationalist instances—Gab quietly deactivated its ActivityPub capabilities, rendering Gab.com into an isolated, centralized, single-node network.[28]

Gab was #fediblocked.

There's an epilogue to this story. In 2022, Donald Trump copied Gab's playbook and built his own alt-right social network, Truth Social, also using Mastodon code.[29] While media outlets reported on the explosive debut of Truth Social—for a moment, it was the #1 downloaded app on Apple's AppStore—one area of the Internet wasn't worried at all: the fediverse. As I observed, Truth was blocked preemptively, with many people using hashtags to coordinate such a block. In fact, Truth never attempted to federate at all. Perhaps Truth learned from Gab's experience and avoided the embarrassment of being preemptively blocked.

## Mastodon and Whiteness

With Gab blocked, white supremacists have been blocked. Problem solved, right?

Not at all. Blocking white supremacists was relatively easy because of their visibility. As Marcia told me, the more difficult racism to contend with is the deadening, repetitive, softer, apologetic forms. "There was a lot of posturing" among white fediverse members, she told me. They would often write posts that appeared at first glance to be supportive, but then turned out to be racist: "I'm a user who's also against racism. *However* . . ." "It's always the caveat," she says.

Marcia's observation echoes what philosopher Myisha Cherry calls "white ignorance," "white skepticism," and "white empathy."[30] White ignorance comes into play often on the fediverse: when a Black, Indigenous, or person of color reports a racist incident, a common reply is for white people to say, "I'm also against racism, but I've been on Mastodon for years and have not experienced this"—the implication being that such racism is rare, or even made up. They then might be skeptical and ask for evidence: "I'm also against racism, but I need proof of what you're saying"—and then when they get proof, they respond with even more skepticism and further demands for evidence. "In order to be convinced," writes Cherry, "white skepticism demands overwhelming evidence. To accept the truth of a hate crime, a skeptic needs to see the videotape. After watching the videotape, he needs to see the whole video because 'context matters.'"[31] Finally, a white empathy response is to shrug off racism: "I'm also against racism, but let's relax—it's only a joke, they didn't mean it, it's just a kid on the internet."

Ro, Marcia, and allies of Playvicious faced constant pushback against their campaign to block racists on the fediverse. "People *hated* me . . . because I talked about the idea of isolating those instances," Ro told me. "Because it clashed with people's juvenile ideas of free speech, you know what I mean?" The call from Playvious members to isolate Gab and similar white Christian nationalist instances was "completely demonized," he said. Ro and his friends faced all three obstacles Cherry identified: arguments that instances like Gab weren't really that bad, constant calls for "receipts" or proof of racist posts, and excuses. It led Ro to conclude that, on the fediverse, "there's more patience and tolerance for bigots than there is for the targets of bigots." He says it's hypocritical. "You want to be open to certain forms of bigotry, but you don't want to talk about how it affects certain people?" The skepticism, demand for evidence, and white empathy began to grind on Ro and Marcia.

If a Black fediverse member gets frustrated with skepticism and white empathy, they might write an angry post about the problems. However, this often results in another form of fediverse racism that Ro identified in our conversation. Ro noted a common response to the concerns of marginalized people—especially people of color—was "just move instances." The idea here is that if a Black fediverse member signs up on an instance that is not doing a good job moderating against racism, they should move to another with an admin who is actively #fediblocking racists and trolls. Taken to the extreme, this argument suggests that a Black fediverse member might be most comfortable on a Black-run instance populated predominantly by Black members.

As Ro put it, that sounds suspiciously like Southern US segregation and "Whites Only" signs. This is "about people not wanting to relinquish their title to this space, you know? Which I think is particularly ugly. . . . People want to say, 'oh, there's no culture [on the fediverse]. It's open for everybody.' No, it's not! It's really not! It's not open for everybody." The "just move instances" line revealed that, instead of wanting to challenge the administration of bad instances, fediverse members merely wanted Black people to change instances constantly. This would effectively hide fediverse racism, because as Black people shifted from instance to instance, everyone else could just keep using the fediverse on their original instance and ignore the plights of Black fediverse members.

Ro notes that the "just move instances" line, coupled with all the harassment that Marcia identifies, leads to people—particularly Black people—giving up and leaving the fediverse. "It's really unfortunate," Ro says. "We're seeing scholars, academics—truly intelligent people that can affect the culture in positive ways have these experiences where it's not even worth it. Because people don't want to see different kinds of faces on the fedi."

These patterns have happened on Mastodon and the fediverse for years. In 2022, during the height of the Elon Musk–provoked migration from Twitter to Mastodon, philosopher of technology Johnathan Flowers observed this pattern and summed it up: "Mastodon is a very white space. It is not unlike other tech spaces where whiteness is predominant. Insofar as this is the case, the norms, the habits, the affordances of the platform will inherit whiteness."[32] For Flowers, these patterns have been so deeply ingrained in Mastodon that it could not be a viable replacement for Twitter, particularly because Twitter, for all its flaws, saw the emergence of #BlackTwitter as a force. Mastodon struggles to do the same.

## The Demise of Playvicious

Facing racist harassment, Ro shut down his Playvicious Mastodon instance. In fact, he did it in 2020—right as the fediverse was staving off the white supremacists of Gab. While the isolation of Gab is often held up as one of the most triumphant moments in fediverse history—and yes, it should be—what often gets lost in this story of triumph is the demise of Playvicious.

Marcia told me that there were "years of buildup" to shutting Playvious down. In one inciting event, she had a stalker who was posting screenshots of her posts to Discord, chat systems, and other instances. She had private "posts that are being screenshotted and shared," they told me. "You have to really be bored or have an obsession to do that." This was just one of many moments when Marcia faced tremendous harassment—she's Afro-Caribbean, she's a moderator making decisions about blocking, they cocreated #fediblock, she's on a Black-run instance. These are multiple targets for harassers to come after them.[33] As Johnathan Flowers puts it, as a moderator, she was subjected "to racist violence over and over and over again to filter out the racists and protect their community."[34] This made them think "I'm not safe here" on Playvicious. "It just got too much for my mental health. Then there's George Floyd [who was murdered by police in 2020] . . . It just weighed on our mental health."

Other interactions took their toll. Ro would call out racism among white trans women and would face bigotry as a response. Marcia, who is currently finishing her PhD thesis on decoloniality, would call out anti-Blackness among Latino/a people and face recriminations for doing so. As American Studies scholar Christina B. Hanhardt notes, one of the most painful aspects of living in our stratified, capitalist societies is internecine struggles among groups who might have more in common than not in terms of marginalization. In societies that overemphasize the individual, distinct identity categories become sites of difference rather than intersectionality, with the result that, for example, wealthy, white queer folks would side with state power in a crackdown against people of color or poor people.[35]

These conflicts can carry over into social media, with the result that people like Ro or Marcia, who call out bigotry among members of otherwise marginalized groups, would face backlash—particularly from white people who observed these interactions and blindly jumped into threads. "They don't understand what's happening," Marcia told me. "To [white people], it just looks like, 'You're saying that a Latina person can be racist, and they're Latina—how can they be racist?'" She would constantly have to explain the complexities of identity: "firstly, you don't know anything about the history of Latin America, you don't understand colonization, despite the fact that you're saying that you do. And you don't understand what anti-Blackness is. You just think it's someone being racist to a Black person without dealing with the nuances." Even with her growing expertise in decolonial theory, they found it tiresome to repeat these conversations many times. "There's a lot of erasure that comes out of that."[36]

It all was too much, so Ro shut Playvicious down.

Overall, for Ro, Mastodon went from the software that helped him climb out of the deep despair of a breakup, out of a lonely apartment in Cancun, and away from Twitter to self-hosted social media, to being merely a vector for white supremacist harassment. When I asked him to reflect on Mastodon, he said,

> I want to be clear. I cannot take away the credit from Eugen [Rochko] for building Mastodon. I think that was an incredible step forward . . . for the

> internet in general. Because it brought the idea of self-hosted social media to the mainstream. And the impact of that cannot be spoken about enough.

Mastodon was ready, Ro notes, to absorb waves of former Twitter users. It is, in many ways, a viable alternative to Twitter/X.

But the demise of Playvicious came about because of obstinacy on the part of Rochko and Mastodon developers, Ro argues. Without good moderation tools, Playvicious's moderators were collapsing under the weight. Despite this, Rochko refused to implement affordances like importable instance blocklists into Mastodon. That safety feature that was called for before and throughout Playvicious's run from 2017 to 2020, but it was only implemented in 2023, years after the demise of Playvicious.

The end of Playvicious seemed to be the end for Ro's respect for the fediverse. While in 2017, he declared "I love the fediverse," seeing its potential to radically reshape the Internet, in 2020, he considered it to be an utter failure.

## Import The-Bad-Space.csv

But then, something unexpected happened. The demise of Playvicious became a mark of shame among fediverse members. I observed people discussing the demise of Playvicious, seeing it as the failure of Mastodon and the fediverse's struggle for a more democratic, equitable, and just digital communications system. For every person who touted victory over Gab, another would point to the closure of Playvicious and say, "Our struggle continues"—or that the fediverse has failed altogether. In fact, Gab is still online, despite being isolated from the fediverse. If Playvicious can't exist, but Gab can, what does that say about the fediverse? Is it living up to its ideals?

Ro heard all this. When I interviewed him in 2022, he noted people are talking "specifically about Black-owned instances on the fediverse, and specifically what happened to Playvicious." Despite the pain of the end of Playvicious, that year he started planning something new: The Bad Space.

The Bad Space is an instance blocklist. It is, per its About page, "an extension of the #fediblock hashtag created by Artist Marcia X with additional support from Ginger to provide a catalog of instances that seek to cause harm and reduce the quality of experience in the fediverse."[37] The website provides a list of instances, provides evidence for why they are on the list, and provides tools for fediverse admins to use to block those instances.

Ro wasn't alone in building a public instance blocklist. Despite Mastodon's refusal to implement instance blocklist importing, other admins decided to publicly share lists of alt-right, fascist, tranmisic, racist, or child exploiting instances. This practice happened off and on throughout Mastodon's history—the short-lived Fediblock.org project was one example—but it started gaining momentum in late 2022 as waves of people moved from Twitter to the fediverse.

A notable project here is Oliphant's List. Oliphant was the admin whose first interaction on the fediverse was being called a "tranny n----- faggot." He realized that if his first experience as a white fediverse member was of being harassed, no doubt many other new fediverse members would experience the same, particularly people of color, and so he went on a mission to build his own blocklist. "The initial experience people had—especially Black people—was: this place is an unmoderated hellhole," he told me. His reaction to such harassment was, "How do I bury you? How do I block you?" He did not see many examples of blocklists, however, so he built his own in 2022.

Recall the arguments against instance blocklists: the *dogmatic free speech* argument, the *open federation* argument, the *grudgelist* argument, and the *admin responsibility* argument. As far as Ro and Oliphant are concerned, the dogmatic free speech argument is just about repressive tolerance: it is code for white supremacist and transmisic entitlement. The open federation argument is now negated, thanks to the net good of #fediblocking instances like Gab—the fediverse is better off since Gab was isolated. This leaves the final two arguments, grudgelist and responsibility.

Could these blocklists devolve into personal grudgelists? Both Ro and Oliphant are very aware of this issue, and both are establishing governance structures to mitigate this risk. In Ro's case, he has partnered with nonprofits, such as the Nivenly Foundation, to help build a governance structure. Currently, Ro is publicly sharing governance documents and inviting comments on them.[38] For his part, Oliphant has joined with a group of admins—the Fedimins—to collectively decide which instances should be on a blocklist. In fact, Oliphant doesn't just offer one blocklist—he has several, tiered blocklists. Each tier indicates a degree of consensus. Tier 0 reflects 60–80% consensus of the Fedimins. Tier 1 is broader, reflecting the instances 50% of the Fedimins agree on blocking. The broadest tier, Tier 3, only requires two admins to agree. Both Ro and Oliphant are seeking out noncentralized governance structures that remove any personal biases they might bring to bear, mitigating against these lists becoming an individual's grudgelist.

Moreover, if a well-moderated instance is somehow inadvertently added to these lists, it's not impossible to be removed. Multiple admins have told me that there are cases of admins who had their instances added to a blocklist and reacted by successfully cleaning up their moderation practices, effectively joining the digital covenant and abiding by the fediverse's ethical standards. In such cases, they have been removed from blocklists. But such cases are rare. For the vast majority of instances added to a blocklist, there's no changing them: they will host racists or transphobes because they believe they have the right to and that such content is an indicator of the freedom of speech. When such instances end up on a blocklist, their reaction tends to be doubling down in the name of free speech instead of moderating against such posts. As Oliphant told me, "You can always tell who a person is when they get blocked."

What about the argument that admins should be responsible for blocking and not offload that task to a third party? Even if governance structures don't negate this argument, then the existence of multiple blocklists does. It may seem as if Ro's The Bad Space and Oliphant's List are redundant. But this is another unique aspect to the fediverse: there is no reason multiple entities can't create and share blocklists for admins to choose from. This is the fediverse—it is noncentralized, and its blocklists can be, too. Indeed, more organizations are producing them, including IFTAS (a nonprofit dedicated to content moderation on the fediverse).[39]

This seeming redundancy is a rejoinder to the "admin responsibility" argument against shared blocklists. As Socrates, the admin of Scholar.social, told me, having multiple instance blocklists requires instance admins to make choices. If an admin wants their instance to be more open, but they still want to block overt racists, importing a smaller blocklist that defederates with racist instances makes sense. If, on the other hand, an admin wants to block more aggressively, importing a bigger list is a better option. "There are absolutely legit reasons why people might want to maintain different blocklists," Socrates told me. "Hell, academia can be obnoxious, so I can imagine people wanting to block Scholar.social. I get it—we're not everyone's cup of tea!" Ultimately, the decision to adopt a blocklist is up to an instance's admin, and with multiple options, admins must do some research to decide which lists to import. As Oliphaunt writes on his page discussing his blocklists, it's "assumed you are a human being with agency, and take responsibility for the blocks on your server, no matter how they get there, either via blocklist import or being entered manually."[40]

However, the biggest argument against the publication of instance blocklists is that, as the benevolent dictator for life, Eugen Rochko does not want Mastodon to have the ability to import them. That argument is no longer valid, either. Earlier in 2023, after the big wave of Twitter users to Mastodon, Rochko acquiesced to those calling for importable blocklists: the feature was implemented in version 4.1.0 of Mastodon. As he told me in an interview, Rochko finds importable instance blocklists "incredibly useful." However, in that interview, he also repeated many of the arguments against their use, cautioning that blocklists are dangerous tools that must be wielded carefully.

Speaking personally, Rochko's acceptance of importable instance blocklists happened just in time for me. Right after version 4.1.0 of Mastodon was released, I started as the admin of AoIR.social, an instance for members of the Association of Internet Researchers. My very first act as admin was to import both Oliphant's Tier 0 list and Ro's Bad Space list. They both came in convenient, comma-separated value (CSV) spreadsheet files, easily downloaded and imported. Thanks to those lists, Aoir.social has enjoyed a very calm existence.[41] But the years of struggle to make such instance blocklist importing possible is not visible in those spreadsheets.[42]

## "Rage, hope, anger, optimism": The Continuing Gifts of the #BlackFediverse

Having easy-to-import blocklists made my life as the AoIR.social admin much easier. Rather than running a social media site wide open to the trolls who might delight in harassing AoIR.social members, we've had a pleasant time sharing cat pictures and talking about the state of the Internet.

There is a danger when something is easy, however. We can forget about struggles to make these "easy" tools. The history of harassment of people on Playvicious can be forgotten as we go about engaging on social media. Instead of knowing the history of how social protocols and digital covenants were developed, the protocol and covenant simply elide the history of racism and bigotry.

Thinking about these complicated histories, I had to ask Ro: if something like The Bad Space and instance blocklist importing were available during Playvicious's run, would Playvicious still be online? "That's a big question, man," he replies. He pauses. Then he says, "The way I think about it now, PV did what it was supposed to do.... It's over." Its demise still angers Ro. But despite all the heartbreak—the demise of Playvicious, the constant attacks—Ro kept giving to the fediverse, spending his time making The Bad Space. He is back with a new Mastodon instance, H-I.social, with a nonprofit organization to support it.

This brought me to another question: Why do this? Why keep giving? "Rage, hope, anger, optimism," Ro answered. These are the gifts of the Black fediverse.

The continued racism—ranging from the overt white supremacist instances to the softer, kinder, and deadening harassment of white skepticism and "choose another instance"—is a cause for rage and anger. In *The Case for Rage*, philosopher Myisha Cherry argues that rage is an appropriate emotion in the struggle against anti-Blackness. Rage has largely been derided in the Western philosophical tradition as an irrational extreme. It violates, for example, the calmness called for by the Stoics. Instead of Stoic calm in the face of injustice, Cherry calls for a particular kind of rage, one she calls "Lordean rage," named after the Black activist Audre Lorde. "Lordean rage," she writes, "is targeted at racism. An organizer who is angry at racial inequality and motivated to end it so that all of us, regardless of skin color, can flourish has Lordean rage."[43]

Ro's Lourdean rage at racism and harassment on the fediverse is palpable. The Bad Space website once featured a large image of a dumpster. On the side of the dumpster, stenciled in black ink, is a bit of poetry: "I thought you were special... I thought you should know."[44] The message is clear: for all those on the fediverse who think they are entitled to harass others, to have their racism heard by all, to drive away people of color, trans folks, Indigenous folks, disabled folks: yeah, you're special. That's why you're tossed in the dumpster that is The Bad Space. Here, the dumpster is the place to dump all that enrages. "There is still a lot of anger about what happened" to Playvicious, Ro told me. "There is a lot of rage.... The Bad Space is one of the culminations" of the painful, rage-inducing history of Playvicious. "I carry that history with me."

The thing is, another palpable emotion I get when I talk to Ro is joy. He is hopeful and optimistic. In his book *Distributed Blackness*, digital culture scholar André Brock argues that the Internet has long been a Black space, and as such is a space of Black joy.[45] There's a popular account on Mastodon, Black Joy, that simply posts pictures of joyful Black people. I fully expect a picture of Ro to appear on this account one day, because he is a joyful person. In spite of all he's experienced, he told me, "I see the massive potential of the fedi." Instead of putting our energy into corporate social media, he envisions an ever-expanding fediverse, where

> we don't have to be regulated by the auspices of profit, right? We can connect in authentic ways, rather than just trying to jury-rig these algorithms to get more attention. Just the potential for not just conversation—imagine the idea of community building! Knowledge sharing! Education! There's so many applications of the fediverse that can be used to improve not just our online communities, but our communities in general. [All of that is] not reliant on anything but us, people like me and you: weird-ass, creative, academics, hobbyists, that have the idea of, "Yo, we just want to have better communities that work for us."

It is hard not to see Ro's return to the fediverse and his contribution of The Bad Space project as a gift back to the network that shunned Playvicious. In spite of it all, Ro is back.[46] "We're in a better place now, but the journey to get here was so arduous."

But in their interview with me, Marcia made yet another key point: the expectation that Black Mastodon members either must be constantly "fist up"—full of Lordean rage—or constantly happy is yet another burden. In contrast to Ro, Marcia has given up hope that the fediverse can be anything special. "Is there hope for the fediverse at large? Meh. I tried. I don't know," she told me. "I don't feel like it's my problem right now." Marcia puts their finger on another problem—the struggle for an antiracist fediverse is not just for Black people, but for all who agree to the digital covenant. To expect otherwise is to commit yet another injustice, placing a burden on people of color while everyone else blithely goes about liking, posting, boosting. And when that happens, Mastodon is not "worth it anymore" for Marcia.

It's hard to blame Marcia for stepping back and withholding their considerable talents from the fediverse. She has done a great amount of work to make the fediverse better, with little reward. However, others are stepping up, struggling to make a Black fediverse. I've observed increasing use of the hashtags #BlackMastodon and #BlackFediverse and people boosting those posts. As André Brock notes, Black media culture is bigger than any Internet site—it has existed for all of Black history. It certainly predates the Internet and social media, and it draws on rage, hope, optimism, and anger. "Black folk use the internet as a space to extol the joys and pains of everyday life," Brock argues, "using its capacity for multimedia expression and networked sociality to craft a digital practice that upends technological beliefs about how information, computers, and communication technologies should be used."[47] In a small way, the struggle for instance blocklists, like Ro's Bad Space, has upended the vision of a totally open fediverse in favor of tools for moderators to strengthen the digital covenant.

In our conversation, Ro demurred at all this. "Yo, man, I'm not trying to save the Internet," he told me. "I'm just trying to get some new ideas out there." But Black rage and joy have the potential to become key parts of the covenantal fediverse—especially because instance admins can now easily and conveniently block white supremacist instances. These are the gifts of Playvicious, #fediblock, and The Bad Space.

# 6
# Paying for It
## The Fediverse's Alternative Economies

"This all sounds great . . . but who pays for it?"

Whenever I tell people about Mastodon and the fediverse, I always get this question. Running social media requires computers. Those computers need to be plugged into a power source. They must be connected to the Internet. All of this costs something, and someone has to pay.

People have received a crash course in the economics of corporate social media over the past few years. We can sign up for TikTok, Twitter/X, Facebook, or Instagram for free. As the aphorism goes, however, "If it's free, then you're the product."[1] In corporate social media, advertisers foot the bill. Corporate social media critics have revealed how this works: in exchange for free access to corporate social media, we spend our time profiling ourselves, declaring our interests, mapping our personal relationships, making posts, and doing the emotional work of communicating with one another, all under the watchful eyes of the corporate social media site.[2] Everything we do is monitored, with the data stored for analysis. Software engineers employed by corporate social media develop algorithms and notification systems designed to keep us on the platform, prompting us to post more and feed those same algorithms.[3]

Fundamentally, the product being sold by these companies is the commodity of user attention. The goal is to deliver user attention to brands and advertisements, who want users to purchase the things they see advertised. A corporate social media site sells "usage data to advertising clients, who in return for paying money get targeted access to users' profiles that become advert spaces."[4] Corporate social media profits when the advertisers pay more than the cost of operating the service. In addition, since users are the ones who create attention-grabbing content for free in the form of posts, likes, shares, and comments, content costs for corporate social media are very low. Harvard professor Shoshana Zuboff provides a succinct label for

*Move Slowly and Build Bridges*. Robert W. Gehl, Oxford University Press. © Oxford University Press 2025.
DOI: 10.1093/9780197776711.003.0007

this: surveillance capitalism.[5] As one blogger put it, using corporate social media is "like being at a big party with all your friends but then realizing that the party is really a Pizza Hut focus group. And also, any pictures you take at the party are owned by the focus group forever."[6]

Surveillance capitalism is lucrative. I recall the awe in 2006 when Google bought YouTube for US$1.65 billion.[7] Much has changed since then. Elon Musk purchased Twitter in 2022 for US$44 billion.[8] Even accounting for inflation, that's seventeen times more than what YouTube sold for. Meta, the owner of Facebook, Instagram, and Threads, saw revenue of US$116.6 billion in 2022—and that was a down year for that corporation.[9] Alphabet, Google's parent company, had revenues of US$282 billion that year.[10]

In contrast, Mastodon and fediverse instances predominantly do not fund themselves through surveillance capitalism. Instead, there are a range of ways to handle server costs. The approaches to who pays and how they pay are as diverse as the fediverse itself: instance admins pay out of pocket, solicit donations, and even form nonprofit organizations.

When I tell someone about Mastodon or the fediverse's approach to funding, that the fediverse isn't controlled by Meta or Google, that it's not driven by an attention-grabbing algorithm, that it doesn't sell our attention to advertisers, and that it is run by ordinary people, they soon ask, "How do they make any money? Who pays for these servers?" When I explain that there are a range of ways that people pay for, and get paid for, engaging with the fediverse, but that none of them replicate surveillance capitalism, most people scoff, suggesting it couldn't possibly work. This may be because there seems to be no other option than surveillance capitalism. People might value community-controlled social media, but when the discussion turns to paying for it, they often retreat to a system that sells our attention to advertising. There seems to be no alternative.[11]

This attitude is in keeping with general perspectives on the inescapability of capitalism itself. Critics of capitalist culture argue that the dominant view is "capitalist realism," where capitalism appears to be everywhere and part of everything with no alternative.[12] In the face of the power of capitalism, people tamp down their ambitions for other economic forms. "Their dreams seemed unrealizable, at least in our lifetimes," write two economists who study noncapitalism. This maintains a vicious cycle: when people give up on finding or enacting alternatives to capitalism, others who might otherwise be interested in life outside capitalism don't even bother trying.[13]

## Yes, There Is an Alternative

The problem with capitalist realism is that it is a defeatist vision. It not only ignores the many ways in which people are engaging in alternative noncapitalist or even anticapitalist practices, it also "overlooks and underplays the central roles that 'non-capitalist' forms of economic organisation perform in everyday life."[14]

Here is yet another facet of the "alternative" in "alternative social media": alternative economies that eschew surveillance capitalism. Much as alternative social media bring about alternative topologies and political structures (noncentralization, covenantal federalism) and alternatives to corporate content moderation (small-scale local moderation, community blocklists), people on the fediverse are building alternative ways to pay for social media and support creative people.

People are using the fediverse in ways that deny corporate social media as the only possible way of socializing online. They are engaging in "community economies," "economic spaces or networks in which relations of interdependence are democratically negotiated by participating individuals and organizations."[15] Instead of relying on the powerful force of (surveillance) capitalism, which seeks to draw everything into itself, many people on Mastodon and the fediverse rely on alternative economics.

This is in keeping with the larger struggle of the fediverse—the struggle for noncentralization and democratic control over social media. This book has showcased activists who have advocated for codes of conduct and the ability to choose with whom to federate. Likewise, the noncapitalist structure of the fediverse is also a form of activism, publicly demonstrating that other ways of running social media are possible. "In an age of 'no alternatives' (to capitalism), drawing attention toward the presence of ordinary anarchist (non-hierarchical, voluntary) forms of organising within society, which are known and familiar to most people, is incredibly powerful when agitating for change."[16] What better way to communicate that noncapitalism is possible than to develop and operate a communication technology in a noncapitalist manner?

The fediverse's alternative economics is most visible in how people fund instances, but it also appears in other practices, as well. Here I will explore server funding models as well as mutual aid practices and how artists are using the fediverse to promote their art. I will also explore the major downside of the alternative economy: burnout.

## Server Costs: Road Trips and Nonprofits

Socrates, the admin of the Mastodon instance Scholar.social, takes an informal approach to funding his instance. "The model that Scholar has always used is, I pay for it out of my pocket, and anyone else who wants to chip in can chip in as well," he told me. He pays for his server out of pocket, but members of Scholar.social can send money to Socrates via Patreon if they wish. "So, right now, with the level of recurring support we've got, we *almost* get to the point . . . where I only have to throw in a small amount of money every few months in order to keep the lights on."[17] Socrates likens his approach to a road trip. "I just want to be like a few people in a car. We split the gas four ways—that sort of deal."

This sort of informal approach is the easiest. Most people start a Mastodon server by providing a personal credit card to a cloud hosting service and paying a monthly fee. The fee ranges based on various options—for small servers, the cost is typically around US$50 per month. Like a driver buying a tank of gas on a road trip, admins might gratefully take a few dollars to offset the cost, but they don't expect it. They're running an instance because they're trying to get somewhere: a place where people can escape corporate social media.

However, much as ad-hoc governance practices on Mastodon have evolved into the more formalized social protocols of codes of conduct, many instances have established formal structures to govern funding. Since the fediverse runs on open technical protocols and open source code, many instances on the fediverse take openness to mean that their accounting has to be open, as well.

CloudIsland.nz, the New Zealand–based Mastodon instance run by Aurynn Shaw, asks members to pay a monthly subscription. Like Scholar.social and many other Mastodon and fediverse servers that ask for donations or subscriptions, Shaw uses Patreon to manage subscriptions. Cloud Island features several tiers of funding, including individual memberships, +1 memberships for two friends, and family memberships. Patreon also allows for members-only blog posts, which Shaw uses to report to the community about the state of the instance, its finances, and poll the membership on governance decisions.

There are other, even more formalized approaches than the Patreon approach used by Cloud Island. Some fediverse instances joined the Open Collective Foundation, a charitable nonprofit organization that operated in multiple countries, providing administrative and fundraising resources for its members. In exchange for taking a portion of donations, Open Collective functioned as an umbrella nonprofit organization, enabling member groups to enjoy that status for tax purposes. Open Collective also provided software that enables members to openly report income and expenditures.

Dozens of Mastodon and fediverse instances joined Open Collective. I talked to Joshua Wood, the admin of Federated.press, a journalism-focused Mastodon instance which has been part of Open Collective. "I have no interest in profiting from Federated.press, so I thought about starting a nonprofit," Wood told me. "But that was a lot of work for a side project." He chose Open Collective not only because of the ability to be part of a nonprofit collective, but also because his journalism-focused instance might need extra protection. "They also offer extended liability insurance for an additional fee, which I opted for," because "you're taking on risk when you operate a platform that hosts content and speech, so my advice is to take all the help you can get." In offering insurance, budgeting tools, and nonprofit status, Open Collective contrasted with the most common channel of funding used by fediverse admins, Patreon, which is a for-profit company that only provides a payment system. As Wood told me, joining Open Collective is "more of a partnership" with a fellow nonprofit organization.

However, I should note that, as of this writing, Open Collective's underlying foundation, the Open Collective Foundation, is shutting down. This leaves the future of partnerships between fediverse instances and the Open Collective uncertain.[18] Given how many fediverse instances relied on Open Collective as a nonprofit umbrella organization, if another, similar organization steps forward, many instance admins might choose to rely on it to gain some formal organizing structure.

Considering the uncertainty of Open Collective–style partnership, some Mastodon and fediverse instances have chosen to fully formalize their structure by incorporating as registered, standalone nonprofits. Chaos.social, a

Mastodon instance with 6,000 active members, is registered in Germany as an *eingetragener Verein* (eV), an association or club, which is very common in German civil society. The admins of Chaos.social receive donations and regularly report on the status of their finances—as of this writing, they have budgeted well enough to have three years' funding on hand for their server costs.[19] Likewise, Ro (the founder of Playvicious.social and the Bad Space blocklist) hosts his new H-I.social instance as part of a US-based nonprofit organization he founded called Hueristic Instruments.[20]

Perhaps the most formalized organization in the Mastodon ecosystem is the Mastodon project itself, which registered in Germany as a *gemeinnützige Gesellschaft mit beschränkter Haftung* (gGmBH, or "nonprofit company with limited liability") in 2021. Mastodon gGmBH trademarked the Mastodon name and logo, licenses the code, pays several developers to maintain the code, and operates the instances Mastodon.social and Mastodon.online. In 2024, Mastodon also formally incorporated as a nonprofit in the United States.[21] On the broader fediverse, other projects are organized similarly. For example, the ActivityPub-enabled video-sharing system PeerTube is developed by Framasoft, a not-for-profit organization based in France.

Whereas a German eV association has minimal reporting requirements, gGmBHs and US nonprofits must offer detailed reports on their financial status. Mastodon is funded entirely by donations and grants. Perhaps unsurprisingly given the attention paid to Mastodon in the wake of Musk's purchase of Twitter, donations to Mastodon gGmBH soared in late 2022, enabling Rochko to hire people, pay consultants for designs, and accept more salary for himself.[22]

With all this money flowing around the fediverse, can it be called a noncapitalist network? As economists who study noncapitalist practices note, just because there's monetary exchange involved on the fediverse doesn't make it a capitalist space. "With respect to the varieties of labour practices: money is not indicative of 'capitalist' relations between 'the buyer' and 'the seller.'"[23] Neither does the fact that many of them offer financial statements, showing income and expenditures, indicate that this is a capitalist system. Noncapitalist economies can include formal reporting and organization. Whether informal or formal, the distinction between the fediverse and corporate social media should be clear: Corporate social media fund themselves largely by selling commodified user attention to marketers and advertisers. Many Mastodon instances rely on donations or subscriptions from members—that is, if they are not simply gifts given from an admin to a membership. They are not turning their members' attention into a commodity to be sold. Server funding on the fediverse is decidedly an alternative to corporate social media.

## Mutual Aid

Another key noncapitalist economic practice on the fediverse is mutual aid. As a scholar of mutual aid defines it, "mutual aid is a form of political participation in which people take responsibility for caring for one another and changing political conditions, not just through symbolic acts or putting pressure on their representatives in government but by actually building new social relations that are more survivable."[24] In a sense, much of the fediverse is mutual aid. When someone runs a server as a service for others—regardless of whether they get donations to offset the costs—it is an act of mutual aid, allowing people to engage in social media outside the surveillance capitalist system. They are not depending on corporate social media to supply this channel of communication, nor are they hoping states regulate corporate social media's excesses. Fediverse instance operators are actively building new social media relations.

Mutual aid is most explicit when people support each other during times of distress. I talked to a South Asian fediverse member who goes by the moniker EdenDestroyer. He actively posts using the #MutualAid hashtag on the fediverse. Working with several others, EdenDestroyer focuses on crowdfunding money to aid trans people who are looking to leave increasingly oppressive, anti-trans Southern US states, such as Florida. He has recruited people on the fediverse who have large followings to boost these posts.[25] "Our work is necessary because we develop reach for people who do not have any reach, and who need it desperately," EdenDestroyer told me. Since Mastodon and the fediverse eschew sorting algorithms in favor of a chronological timeline, EdenDestroyer's strategy of using hashtags and asking for boosts is the most effective way for these posts to spread across the network.

The money raised through these campaigns helps trans folks pay their bills, look for jobs, and pay moving expenses to escape anti-trans environments. As EdenDestroyer told me, "The 'mutual' part of the 'aid' signifies the mutually agreed upon unconditional relationship between the donor and the aid asker." The unconditional aspect is important. The slogan of mutual aid is "solidarity, not charity." According to mutual aid scholars, this is because many charities pathologize the individuals who are being helped, demanding that in exchange for support, the recipients of charity conform to mainstream social values.[26] In contrast, the approach taken by EdenDestroyer and his colleagues does not pathologize victims or demand that they change. Instead, pro-trans mutual aid posts focus on state-level oppression of the people being supported, not on some perceived failure of the people who live in those states. This reflects the mutual aid principle of "getting support in a context that sees the systems, not the people suffering in them, as the problem can help combat the isolation and stigma."[27]

One person who is also active in #MutualAid campaigns is vanta rainbow black. She is a trans woman who herself has benefited from financial support from others, including during their own move from rural Florida to Seattle, Washington. In between their posts about her pirate radio station or her selfies modeling her handcrafted, black heart clothing, she boosts EdenDestroyer's mutual aid posts. vanta finds the fediverse far more amenable to these noncapitalist posts. "On Facebook, most people would just call you a liar, or scammer, or some shit, if you tried posting a link to any sorta donation thingy," she texted me via Signal messenger. In contrast, "on the fediverse, mutual aid posts are largely respected. People make posts with their donation links and such all the time, and I boost every single one I can find."

This is not to say that no one on the fediverse harasses vanta for sharing mutual aid posts. But this is where blocking and defederation come into play. "What happens when people are shitty about it is, they get reported and put on #fediblock," vanta texted me. "When instances ban people for making donation posts, those instances get widely defederated or pressured to change their rules. I've seen it happen quite a few times, lol!"

While vanta boosts mutual aid posts of others, she also asks for help, as well. Recently, she posted to the fediverse, asking for help in finding a job in Seattle, expressing their vulnerability as someone who is underemployed. Moments of vulnerability like this are common on the fediverse. They are symptomatic of noncapitalism: rather than projecting dominance and success (as is often the case among corporate social media influencers), fediverse members use vulnerability to build solidarity. As scholars of mutual aid argue, "being vulnerable, then, is not to be a passive recipient of aid, someone who needs a bit of free food in order to be an employable subject one again. It is a state of social possibility; of latent connectivity with others to create more just ways of organising communities."[28] Being openly vulnerable is a potent repudiation of capitalism, the system so many of us imagine is the only one possible. As James Baldwin puts it, "one can give nothing whatever without risking oneself. . . . If one cannot risk oneself, then one is incapable of giving."[29]

vanta can risk herself and be vulnerable because their needs are not pathologized, and she can give back to the fediverse because her contributions are valued. As I will show in a later chapter, vanta will give back in a major way by organizing the #fedipact, an agreement among many fediverse instance admins to preemptively block Meta's Threads as it appears on the ActivityPub network.

Like noncapitalist forms of server funding, fediverse-based mutual aid "exposes the failures of the current system and shows an alternative. It builds faith in people power and fights the demobilizing impacts of individualism and hopelessness-induced apathy."[30] Fediverse-based mutual aid does not demand that everyone get a job and stop asking for money. It does not blame people for their circumstances. Instead, people like EdenDestroyer and vanta black are building a noncapitalist system of funding and support for one another, an "alternative, politically motivated ideology of collective responsibility, co-operation and mutual survival."[31]

## Creative Economies

People have used corporate social media to promote themselves. This includes artists, writers, and musicians. They post notices about their work and engage with their fans. While this is a form of advertising, it is not the same as online behavioral advertising—it doesn't rely on tracking people across the Internet, and it is on a personal, nonindustrial scale. It's a form of self-promotion that artists rely on to ply their trade.

Like everyone else, though, artists' promotional posts, as well as their creativity, are appropriated by corporate social media as freely given, attention-grabbing content.[32] Artists see this as part of the deal—they make content that Facebook, Instagram, or Twitter/X uses to attract audiences, and the artists in turn can reach that audience.[33] Can artists shift to the fediverse and make a living instead of being subjected to surveillance capitalism?

The answer is yes, but artists are also hedging their bets.

I spoke with David Revoy, a French digital artist who began his career doing portrait art in public on the streets of Avignon, France, in the 1990s.

He transformed into a digital artist in the early 2000s, favoring open-source tools like Krita and Blender. Since 2014, he has been creating a popular webcomic series, *Pepper&Carrot*, a fantasy about the adventures of a young witch and her cat.[34] The comic has been adapted into an animated show and is translated into over sixty languages.[35]

Revoy's funding model is to crowdfund each issue of *Pepper&Carrot*, and to do that, he has relied on working in public on social media, posting frequently and sharing videos about his work. He sees this as replicating the street art approach of his Avignon days. "I really started as a street artist. . . . It's a bit fundamental in the way that I work," he told me. "I still have this feeling of public space." Much like donation-driven Mastodon instances, Revoy seeks donations to support his art. "It's not selling my service—it's more like what all other artists in Avignon do. Doing a spectacle, and showing the art for the donation process."

Initially, in the mid- to late 2000s, he used Twitter and Facebook for his public art method, sharing drawings and watercolors. However, as he told me, Twitter seemed to suppress his posts—to have his work seen, he felt he had to play to an opaque algorithm. "I began to feel like the game was deeply cheated on Twitter. That I could make all the effort to put the best artwork, to involve myself with the community" but still gain little traction in terms of followers or exposure. He also increasingly saw art on Twitter as being dominated by celebrating the latest Hollywood movie or video game. It "felt like on Twitter like artists [were] getting inspired by new, industrial movies or something, like a new game by Nintendo." He suspects—but can't confirm—that Twitter was using its algorithm to boost these posts for promotional purposes. Revoy detests this celebration of cultural commodities, he told me.

In 2017, some friends suggested he join Mastodon. When I asked him to compare being an artist on Twitter to being an artist on Mastodon, he simply said: "Mastodon: wow!" He says that Mastodon's chronological, nonalgorithmic ordering approach means his posts are much more likely to be seen by his followers instead of being buried by posts celebrating the latest cultural commodity. He told his Twitter followers that they would get the same posts on Mastodon as they would on Twitter, and so many of them followed him to the fediverse. His follower count grew steadily, ensuring that the former street artist could have a large online public to create for. "With Mastodon, you get a fair opportunity to get your artwork in a sort of public place," he told me. He eventually closed his Twitter account and now relies primarily on Mastodon for his street art–style public promotion of *Pepper&Carrot*.

Revoy's art is also notable because he licenses much of it, including *Pepper&Carrot*, with permissive Creative Commons licenses, meaning others can reuse or modify his work without getting his explicit permission.[36] While Revoy does sell his art, he also contributes it to a commons, where other artists can build on Revoy's work. For example, voice artists and animators have revised *Pepper&Carrot* into an animated cartoon—and Revoy supported their choice to crowdfund that project, even though he did not receive money for it.[37] Revoy's use of Creative Commons reflects the ethos of much of the fediverse, which itself is built on permissive licenses, allowing people to use and modify its protocols and software.[38] It also reflects a long-standing shift in digital media production, where the line between artist and audience is increasingly blurred.[39]

I also spoke to Jessica Rowbottom, a UK-based LGBTQ+ cabaret singer who performs a show called The Bleeding Obvious. In The Bleeding Obvious, Rowbottom intersperses piano-and-voice songs with anecdotes and comedy focusing on growing up queer in the United Kingdom.

Like many people, she ultimately stopped using Twitter after Musk's takeover. "I'd been using Twitter extensively as a performer—less so personally, but certainly as a performer," she told me. "I performed at [the Edinburgh Fringe Festival] in August, 2022, and a substantial amount of my marketing and promotional work of getting into people's heads who were going to that particular festival" took place on Twitter. When Musk bought Twitter, she essentially stopped posting there.

But "losing Twitter was a massive thing for people, for a lot of us [performers]," she said. Her fellow performers have

> certainly been seeing a massive difference as to how people are discovering shows without Twitter. I think that's because of the algorithm, the Twitter algorithm, would put stuff in front of you. If you hashtagged something with #fringe, then it would shove it in front of other people, even if they haven't specifically requested it.

Their experience replicates what many people have found on corporate social media: if someone is lucky enough to be graced by the content algorithm, they can get their posts promoted to wider and wider audiences. (For people like David Revoy, who said he wasn't blessed by the algorithm, the feeling is of shouting into a void.)[40]

Musk's Twitter takeover prompted Rowbottom to shift to the fediverse. She decided to self-host her own Mastodon server, blop.social. As a queer performer, she chose self-hosting for safety purposes—she didn't want to be "at the whims of another Elon Musk" by trusting any other server to keep transmisic people away. In addition to doing all the work of running her own server, she also found the fediverse to be demanding for self-promotion. "Twitter is low-effort—or was," she said. "The fediverse is effort. . . . You've got to go out and look and find your audience. . . . You're building your own little hub system." She likens this to pre-corporate social media promotional techniques: hanging up posters, calling people, telling people about upcoming performances. She recommends the fediverse to fellow performers—she just warns them that it is hard work.

However, Rowbottom also hedges her bets by posting to Facebook, Instagram, YouTube, and TikTok. Even despite the opacity and constant changes to sorting algorithms in corporate social media, she said, "getting stuff out there via organic reach is still a lot easier on Instagram and Facebook than it is on Mastodon and the wider fediverse in general" because the content may be algorithmically promoted to audience members. She does all the promotional work herself—"graphic design, duplication, getting stuff out there, driving around the country, making sure posters are put in the right place"—and so she's willing to use multiple channels if it makes her promotional efforts go farther.

Indeed, Rowbottom is not alone in using multiple channels. Even though his main channel is Mastodon, David Revoy also maintains Instagram, Facebook, and YouTube accounts. Other artists I've talked to or observed do something similar. While it is possible to shift to the fediverse and make a living as an artist, the lure of promotional algorithms—predominantly designed to keep attention on corporate social media—keeps many artists active on corporate social media, even despite their many problems.

At least for creative work, resistance to surveillance capitalism (if not all of capitalism) remains difficult. This is not surprising, given that so much of today's hustle culture of influencers, gig workers, and entrepreneurs using corporate social media reflects the underlying precarity of contemporary life in capitalism—a precarity that artists, who often do not have formal employment, understand and begrudgingly accept.[41] Corporate social media has been built by profiting from the free labor of its users who create and pay attention to its content. Those who give their labor for free within digital media can hope for hypervalorization—that is, stardom and outsized rewards.[42] They hope to be graced by algorithms so they can convert large audiences into cash.[43] This is in large part what makes corporate social media so sticky for people who do precarious jobs.

But by shifting much of their focus to the fediverse, which does not rely on behavioral advertising, Rowbottom and Revoy have found that the fediverse has removed one intermediary between them and their audiences: content-shaping algorithms. Using the fediverse means that artists are not subject to the political economy of surveillance capitalism, which shapes audience taste through algorithmic boosting or deemphasis.[44] Since fediverse instance operators have no interest in gathering data on who likes what and who is connected to whom nor interest in selling member attention to advertisers, there is no reason to use algorithms to boost particular posts or accounts (and, conversely, reduce the reach of others). This at least starts the process of challenging the dominance of corporate social media in the lives of creative people and creates some alternative space for artists to reach audiences online.[45]

While such algorithms worked for Rowbottom—she found it easier to reach people on Twitter or Instagram—they didn't work for Revoy. Instead of offloading decisions about what is valuable to an algorithm, the fediverse simply relies on individual members to boost what they like. In the absence of an intervening, surveillance capitalist algorithm, artists like Rowbottom and Revoy thus have a chance to more directly connect with their audiences on the fediverse, who in turn have more channels to directly support them outside of the capitalist cultural industry.

## The Cost of Noncapitalism: Burnout

The fediverse provides a key way to escape surveillance capitalism. Instead of being the product, alternative social media allows members—or, to put it more sharply, *requires* members—to be producers. Starting and funding servers through donations, boosting mutual aid posts, or promoting art for an audience instead of an algorithm takes a great deal of human effort.

While that work can be fulfilling, it can also take a toll. Asking for donations, managing servers, and selling art or tickets can grind people down. Donations can be an unreliable channel for server funding. Federated. Press's Joshua Wood told me, "It's really hard to actually get people to donate. I don't think that's unique to us. It's something I've heard from most of the other admins I've talked to, and it's a frequent topic in the fediverse." This difficulty is compounded when umbrella organizations, such as Open Collective or Patreon, change policies or even shut down altogether.[46] And while mutual aid posts are much more respected on the fediverse, they still can attract trolls who harass people asking for help. For musicians and artists who often do all their own promotion and booking, hustling on the fediverse is a tough gig. As Rowbottom told me, "You have to put more legwork in, and it's difficult for the one-man bands now because of that."

Adding to the anxiety, some of the most important work on the fediverse is largely uncompensated. As is well-documented by researchers and confirmed through many interviews I've conducted, community moderation is also a difficult, labor-intensive practice.[47] Even with shared ethical values in a covenantal fediverse, and even on small community instances, moderation takes a great deal of difficult, emotional work. This work is often unpaid. In my observation, and based on interviews with admins and moderators, only a tiny fraction of fediverse moderators receive any payments. Socrates, the admin of Scholar.social, reported paying moderators on a case-by-case basis whenever he faced a tough decision and needed to consult with others. Eugen Rochko noted that moderators on Mastodon.social and Mastodon.online receive payments. But overall, the vast majority of fediverse content moderation is done by unpaid volunteers.

Mawr, the admin of a Mastodon instance called Plush.City, talked to me at length about the work of moderating an instance. Every day, Mawr starts by checking #fediblock, noting what instances or accounts are mentioned there. They also check the local timeline. She gets user reports and checks them against the very extensive blocklist on Plush.City. The bulk of the work is responding to reports. "To do it right requires communication," they told me. But "you have no admin tools to let you communicate with other instance admins, or ways of keeping communication around an incident internal to that incident. You have to do all that manually. That takes time, organization, and a lot of spoons."[48] When there are internal conflicts between plush.city members, Mawr must do the very difficult, time-consuming work of resolving them. Such work takes a toll on people like Mawr.

Another area in which people predominantly volunteer their work is creating and maintaining the code that runs the fediverse. The free and open-source software (FOSS) code for Mastodon has involved the work of hundreds of people, and the vast majority of them are unpaid. As Rochko told me in our interview, it's only been recently that the project has had enough donations to pay developers, and those developers are only a tiny fraction of the total number of people who contribute code. Mastodon's history as a FOSS project is marked by moments when contributors argued they were not getting enough credit, some of whom have stopped contributing.[49] It is also marked by demands that Mastodon implement this or that feature—features that can be contradictory—requiring its developers to make decisions that inevitably displease many of the software's users. Mastodon joins the ranks of FOSS software projects that have seen people burn out because their contributions weren't recognized or they couldn't continue to handle the demands of their users.[50]

While many fediverse members engage in mutual aid and support one another, while server costs can be offset, and while artists might be able to promote themselves to audiences, the specter of burnout constantly haunts the fediverse. Such burnout has led to the closure of several Mastodon instances or moderators stepping back from active involvement (think back to Marcia X in a previous chapter). Burnout is even felt by someone who is one of the most financially well-supported, Eugen Rochko himself. "I don't have a lot of reference points for when burnout begins and how to identify it," he told me. "But if I had to just say, like, from my feelings, I would say that I have been struggling with burnout for the last two or three years, and still I'm pushing through it." Since Mastodon gGmBH is both a FOSS development project and a manager of two large Mastodon instances, the stress on people like Rochko and the few other employees is tremendous. They must both attend to code development, deciding on which features to implement and how, while moderating instances with tens of thousands of people.[51]

Exacerbating his feeling of burnout, the existence of surveillance capitalist, for-profit social media gives Rochko a vision for how things could have been otherwise. He compares the financial rewards people like Facebook founder Mark Zuckerberg or Twitter cofounder Jack Dorsey have enjoyed for surveillance capitalism to the relatively meager donation funding the Mastodon gGmBH organization has received. "The people that I compete with, the people who own the competitors—Twitter, Facebook, and so on—they're billionaires. They have so many resources. And they benefit personally from everything they do with the platforms." Rochko does not enjoy such rewards, even after the recent growth of Mastodon.

However, while venture capitalists have repeatedly offered large amounts of money to the Mastodon project—particularly after Musk bought Twitter and there was a massive shift away from X—Rochko has refused to take the funding.[52] Such a move would violate the core ethic of the project—it is nonsurveillance capitalist. Venture capital investors demand growth in revenues, and the dominant method of making money is selling user attention to marketers. Even Rochko's most ardent critics express admiration for his refusal to accept venture capital.

As Rochko told me, despite burnout and the lack of financial rewards relative to corporate social media, "I'm still committed to working on this, because Mastodon means a lot to me. It is my creation, and I use it every day. It is the platform I use to get news, to get cat pictures, to post my own photography, for fun times."

## So, Who Pays?

Who pays for the fediverse? The answer is: all of those who create and participate in it. They donate, run instances, spend time moderating, contribute code, post, write codes of conduct, and read codes of conduct. Ordinary people discuss the wisdom of limiting or blocking other instances. They make connections to one another, share art, and buy art. They get into disputes over posts, fret over the future of the fediverse, ask for donations, and give donations. They shitpost, share memes, despair over the state of the world, and laugh with each other. They burn out and return.

Ordinary people do the work of making the covenantal fediverse run. Because this work is largely noncapitalist, one of the ethical values of the covenant is a rejection of surveillance capitalism. None of this work is being bought or sold by surveillance capitalist corporations. Despite capitalism and the defeatist idea that everything must be commodified to exist, the covenantal fediverse stands as an alternative.

While burnout is a possible outcome of social media outside surveillance capitalism, the fediverse's noncapitalist approach is viable. Instances such as Mastodon.social, Scholar.social, and CloudIsland.nz have been on the ActivityPub-based fediverse since its earliest days and are (as of this writing) still going. When instances shut down, their members can move to other servers. Mutual aid posts are now accepted on the covenantal fediverse. Artists continue to share their work.

All of this noncapitalist creativity and activity is sure to draw attention. In fact, none other than Meta, the owner of Facebook and Instagram and the world's largest social media corporation, is now experimenting with ActivityPub-based federation. Can the fediverse remain noncapitalist after Meta shows up? I will turn to that question soon.

But first, there is one material and economic aspect of the fediverse that I must address: the role of digital media in the climate crisis.

# 7
# To Finity and Before
## Environmentalist Experiments on the Fediverse

What do we expect to find at the end of the infinite scroll? You've seen people doing this—swiping up on their phones while on corporate social media sites Twitter/X, TikTok, Instagram, or YouTube shorts. Maybe you've done it, too. You get through a post, you scroll down, there's another, then another, then another.

Swipe, swipe, swipe. What's at the end? With billions of people posting content to corporate social media, it seems that there is no end to the infinite scroll. As researchers report, users feel "caught in a loop" as they passively scroll through content—this is another attention-demanding aspect of corporate social media.[1]

Corporate social media are invested in the scroll appearing to be infinite. They operate on a logic of infinite growth. Meta, the parent company of Instagram, Facebook, and Whatsapp, triumphantly reported growth in monthly active users to its investors in 2022.[2] According to *Forbes,* Instagram is poised to surpass TikTok in 2023, which itself has seen its userbase grow rapidly in the past few years.[3] Not to be outdone, Google reports that YouTube Shorts (their answer to TikTok videos) is rapidly growing.[4]

That's a lot of content to scroll through.

Growth, it seems, is an unadulterated good. Economists who study the life cycle of corporations argue there is a "growth imperative," where, to put it bluntly, the organizations must endlessly grow or they will die.[5] Corporate social media are no different. The infinite scroll becomes a synecdoche for capitalism: endless expansion, weightless media, and boundless energy.

Infinite growth is an impossible fantasy, however. We live in a finite, material universe. There must be an end to the infinite scroll. All posts on social media must be physically stored somewhere. Corporate social media users are invited to think that our data lives on an ephemeral "cloud," but that only hides materiality: "the cloud" is a shorthand for server farms—collections of thousands of computers.[6] The infinite scroll of corporate social media

*Move Slowly and Build Bridges*. Robert W. Gehl, Oxford University Press. © Oxford University Press 2025.
DOI: 10.1093/9780197776711.003.0008

relies on these server farms, which are comprised of material "coils of coaxial cables, fiber optic tubes, cellular towers, air conditioners, power distribution units, transformers, water pipes, computer servers . . . flows of electricity, water, air, heat, metals, minerals, and rare earth elements."[7] Compounding the problem, these huge server farms, which often are water-cooled, have been built in desert environments, including the Western United States, which has been struggling through a water crisis.[8]

The devices people use to infinitely scroll are also material. Smartphones and many computers run on batteries, which rely on lithium, cobalt, and nickel mined out of the ground. The plastics used in their construction come from fossil fuels. While device manufacturers pump out new versions of laptops, phones, and tablets every year, inviting consumers to toss out their current devices in favor of the new, the underlying resource infrastructure manufacturers rely on is rapidly dwindling. The devices we toss out often end up in e-waste dumps, with impoverished people burning them to extract some of their raw materials—poisoning themselves with toxic gasses in the process.[9]

We're also at the point where e-waste is littering space. According to the European Space Agency, there are an estimated 130 million space debris objects orbiting above us.[10] Given that much of that material in space has to do with communications infrastructure, our infinite scroll is also filling up the very skies above our heads.

Taken as a whole, the infinite scroll is not just a synecdoche for capitalism; it's a synecdoche for the climate crisis and environmental injustice. Because of the climate crisis, which is causing more and more weather disasters from flooding to hurricanes to wildfires, we must consider the role of digital media. The climate-altering properties of digital media are "an urgent problem. Effectively addressing them will mean disrupting the aesthetic and operational norms of some of the largest corporations on earth."[11] These corporations produce the slick, shiny interfaces of the infinite scroll to distract us from "violent infrastructural practices of material extraction at the very foundation of technological infrastructure and its sustained maintenance."[12]

There must be an end to the infinite scroll. And there must be something better once we put it to its end.

Since the fediverse is an open system, one run predominantly in a noncapitalist manner, it is possible fediverse members can shape it in new ways, allowing them to engage in social media while reducing the impact of digital media on the climate. Two related groups of cutting-edge thinkers—degrowth thinkers and solarpunks—are experimenting with the fediverse now, looking to mitigate how digital media contribute to the climate disaster. These academics, activists, and artists are taking advantage of the fact that the fediverse is under ordinary people's control and are exploring ways to radically reduce its impact on climate emissions. Taken together, degrowth and solarpunk provide multiple ways to counter the fantasy of the infinite scroll.

## Degrowth

Aral Balkan is one half of the Small Technology Foundation. Like the name implies, Small Tech is opposed to Big Tech's obsession with growth.

As a child growing up in Malaysia and Turkey in the 1980s, Balkan was lucky: he grew up with a personal computer. It was an IBM PC, and it was *his*, he told me. He learned the coding language BASIC on it, he played games, and he made his own games. His power over the PC was so complete that he accidentally burned it to the ground by installing an incompatible hardware device. (He has since learned his lesson, he told me, and has not set fires to any other computers.)

In our conversation, he argued that, since his childhood, computation has shifted away from our control over personal devices and toward distant, monopolistic centers—including corporate social media. Instead of having control over our own computers, Balkan says we have ceded that control to centralized systems whose definition of success is endless growth. "What we're doing is incentivizing the growth of these centers," he told me. "That's how we end up with the monopolies that we have today—the silos, the Facebooks, the Twitters, the Snapchats." These companies promote a fantasy of infinity. "The success criteria for big tech is: how can this entity get as large as it can and keep growing at infinite scale?" For Balkan, who recognizes the environmental impact of corporate social media, such growth is a "euphemism for the extinction" of the world.

It's time, Balkan and the Small Tech Foundation argue, to degrow things like corporate social media. Degrowth is a way of thinking that counters years of capitalist fantasies of infinite growth. Writing in *The Economist*, Susan Paulson states that the objectives of degrowth are "to reduce the quantity of material and energy that wealthy economies use, to curb cultural and personal obsessions with growth, and to reorient societies around care and equitable well-being."[13] Degrowth economists have identified several industries that are particularly wasteful, including fossil fuels, meat and dairy, fast fashion, aviation, and advertising.[14]

Advertising—specifically behavioral advertising—is central to corporate social media. Harvard professor Shoshana Zuboff popularized "surveillance capitalism" as a label for behavioral advertising.[15] Balkan uses a more brutal term: people farming. Like any cattle, he argues we are farmed by corporate social media. And such farming is wasteful. "Facebook expends so much energy farming you, and analyzing you, and trying to manipulate you, and running its global ad system . . . that's wasting energy," Balkan told me. "Well, it's not wasting it, as far as they're concerned, because they're making money. But it's wasting it as far as you're concerned, because you just want to have a conversation with someone." We believe we're having a simple conversation through corporate social media, he argues, but behind the screen is a complex and energy-intensive surveillance machine dedicated to selling user data to marketers.

If eradicating the wasteful industry of behavioral advertising is a degrowth principle, then the fediverse might well already qualify as a degrowth medium. In keeping with their noncapitalist approach, fediverse software developers, admins, and members have explicitly refused to implement behavioral advertising technologies. This is why Balkan has actively engaged with the fediverse. Social media that does not use behavioral advertising is necessarily less energy-intensive.

Admins who install fediverse software, such as Mastodon, Lemmy, Pixelfed, or PeerTube, install tools to facilitate conversations, including interfaces for members to speak to one another, moderation tools to help moderators enforce their codes of conduct, and algorithms to help their instances federate. What is not installed are all the other accouterments of people farming: software to collect what instance members are saying and doing and algorithms to pick apart what they are saying to build consumer profiles of them. Most importantly, admins installing fediverse technologies won't install the true user interface of surveillance capitalism: the bidding systems through which marketers and advertisers can purchase instance members' attention.[16]

There is more inherent degrowth on the fediverse than just refusing behavioral advertising. Many fediverse instance admins do something that may seem somewhat bewildering: they lock down their registration pages. They *don't want to grow.* Some admins allow new members to sign up by applying for a spot—that way, they can vet would-be members. Others won't ever allow anyone to sign up. Called "instances of one," their sole member is the person who set them up. As a "small is beautiful" advocate, it's not surprising that Aral Balkan runs such an instance. And while his instance only consists of him, he still can reach many people on the fediverse: he follows almost 9,000 other accounts and has over 40,000 followers.

Limited sign-ups on corporate social media seem almost unimaginable. Corporate social media are often publicly traded companies, beholden to their investors, who demand growth quarter after quarter. Investor-driven thinking leads to what philosopher Bernard Stiegler calls "short circuiting": the overriding emphasis on short-term profits.[17] For corporate social media, this means finding more and more people to sign up, intensifying their engagement—or simply spinning up new services to gain more users.[18]

In contrast, on the fediverse, staying small is a common practice. It helps instance admins in many ways: it makes moderation easier and keeps maintenance costs down. Multiple admins told me they keep their instances small so they could be more easily moderated, citing the labor intensity of moderation at larger scales. If an instance stays relatively small and moderates well, it's far more likely to be widely federated.

Right out of the box, the way people are using the fediverse holds great potential to degrow the growth logics long associated with social media. However, there is one area in which the fediverse is often in line with the same energy-intensive practices of its corporate counterparts: its reliance on the cloud.

## Clouds Are Hard to Dissipate

During that influx of former Twitter users fleeing during the Elon Musk takeover, when hundreds of thousands of people signed up for Mastodon in late 2022, I heard the same thing from many instance admins: we must scale up our servers. We need more CPUs, more RAM, more storage. Cloud providers were more than happy to oblige.

A 2019 research paper illustrates how this works. A majority of fediverse instances run on commercial cloud services, such as Amazon's AWS, Cloudflare, Sakura, Digital Ocean, or OVH.[19] The benefits of using commercial cloud services are many: they tend to be easy to set up, with prepackaged, one-click options. (I built my first Mastodon instance, an instance-of-one, using a Digital Ocean one-click process.) Instead of maintaining server hardware locally, the cloud service does the job. If an admin faces a massive uptick in traffic, the cloud service allows for easy upgrading of hardware. This is precisely what Mastodon and other fediverse admins require during waves of new members.

There are four shortcomings of concentrated use of Amazon and Cloudflare. First, it is a major flaw in the attempt to make the fediverse noncentralized. While instances are run by admins in a noncentralized manner, from the perspective of where they are physically hosted, the fediverse is far more centralized than not. A map found at the Fediverse Observer shows inordinate concentration of instances in major cities. This is not just a function of the cities having more people willing to try out the fediverse; it's a function of where data centers are located.[20]

Second, the use of Amazon and Cloudflare could also be a violation of the ethical covenant many fediverse instances adhere to, since these are corporations with poor reputations for their antiunion activities or willingness to host conspiracy theorists.[21] Related to this, a third shortcoming is that the heavy use of cloud services grates against the noncapitalist practices of the fediverse. Most cloud services are massive, for-profit enterprises, and when fediverse admins pay their cloud service bills with donated money, they are transferring those donations out of the noncapitalist fediverse into a decidedly for-profit sector.

Fourth, and most relevant here, the fediverse's use of cloud services simply replicates the same problem posed by server farms in general: their environmental impact is tremendous. If the fediverse merely repeats the corporate social media server farm structure, it could amplify the ecological problems wrought by corporate social media. Setting up the fediverse in this way would work against degrowth and only contribute to the climate disaster.

Part of the problem stems from a now unquestioned assumption we have about cloud services: they are always available. Another degrowth thinker, Roel Roscam Abbing, a Dutch PhD student working on a dissertation about the fediverse, knows what's to blame for the always-on mentality: the number 9. Specifically, among network infrastructure engineers, the more 9s, the better. A network that's available 99.9% of the time is better than one only up 99% of the time.[22] Better still is one that's up 99.99% of the time. All it takes is more resources—more energy sources, more redundant computer server infrastructure, more cooling capacity. This all sounds simple enough, but very soon network services run into a sort of Zeno's Paradox of Reliability: each subsequent 9 requires exponentially more and more resources. As Roscam Abbing writes, "availability is not only predicated on energy availability but also the availability of other system components. Therefore it will also require a redundancy of upstream network connections, server hardware and additional monitoring infrastructure."[23]

Facebook and Twitter want users to be able to log on and infinitely scroll and post at any time. To achieve such reliability, however, requires massive amounts of energy and hardware—they regularly cycle through hardware, draw power off energy grids, and use water to keep their systems cool. Cloud storage facilities might seek to use renewable resources, contracting with power companies to use wind or solar power. However, the quest for 9s requires them to have a fallback. "To guard against a power failure, [server farms] further rely on banks of generators that emit polluting diesel exhaust."[24] In fact, some of the biggest manufacturers of data center generators are Cat and Cummings, companies that normally invoke visions of big, diesel-smoke belching machines, not light and airy digital media apps.

Ultra-high reliability is why the Twitter wave prompted many admins to quickly scale up their cloud servers. Twenty years of corporate social media have conditioned people to expect an always-on experience—even from alternatives, like Mastodon and the fediverse. As people move from services like Twitter to the fediverse, they expect whatever instance they join to be reliably on, all day long. When choosing between running a server in one's home—where one can't guarantee the power will always be on, or the computer will always function—versus running a social media service on Amazon, many instance admins choose the latter. This is a recipe for more centralization in cloud services—a direct repudiation of the fediverse's goal of noncentralization.

Degrowth thinkers like Balkan and Roscam Abbing have rigorously diagnosed the problems of corporate social media: the use of behavioral advertising, the obsession with growth, and the always-on mentality. While people on the fediverse might be resistant to behavioral advertising, and while many fediverse admins resist growth for growth's sake, the fediverse does rely on the cloud to be always on.

Degrowing federated social media will require cutting through clouds and imagining futures past the current way of thinking, finding alternatives to the dominant, corporate social media approach. Here, fediverse members turn to science fiction.

## Solarpunks

In the face of the climate disaster—in the face of more intense storms, lengthened wildfire seasons, droughts, climate refugees, and rising seas—grief and despair are common responses.[25] Hope seems to be in short supply. This is where solarpunk shines. "We're *solarpunks* because the only other options are denial or despair," declares one manifesto.[26] Solarpunks can be found on the fediverse at Mastodon instances such as Sunbeam.city and Solarpunk.moe and the Lemmy instance SLRPNK.net, whose members share news and ideas about futures beyond the climate crisis.

Solarpunk originated in Brazilian science fiction in the early 2010s.[27] Unlike cyberpunk or other punk science fiction genres, solarpunk is not dystopian. Appearing in anthologies, solarpunk stories feature people shedding unnecessary possessions, using tree root networks to communicate, or teaching children new ways of relating to the land.[28] Solarpunk is marked by people doing the hard work of rebuilding the world. They are not analogues for contemporary people with climate grief. They are future selves who don't mourn—they organize to overcome the ecological damage we are wreaking now. They are our best future selves. "The work is never finished," says one character. "But at least it is begun."[29]

As should be clear, this genre of science fiction is "an uprising of hope against the daily despair that these times bring."[30] It's unabashedly about *porvir otimista*—an optimistic future where inventors and activists work together to create independent, sustainable communities that aren't bent to the will of global capitalism.[31] As fediverse member and software artist Marloes de Valk told me, we need to live "post-depair," "knowing and understanding the complexity and depth of the trouble we are facing (climate crisis and interlinked oppressions and injustices), thus knowing how unlikely it is that the tide can be turned quickly enough, yet refusing to do nothing."

I spoke with T. X. Watson, the author of a solarpunk short story and former editor of a solarpunk webzine. For Watson, solarpunk is speculation about the near future. Instead of envisioning a far-off future (and thus skipping the revolution), Watson's approach to solarpunk is to ask: "What do people who are alive now need to be envisioning as part of the choices they will make in their lives in order to be in progress toward a world we would like to live in?" What steps are possible now to mitigate the climate disaster and bring about justice?

Watson's story, "The Boston Hearth Project," illustrates their perspective well.[32] In the story, a group of activists commandeer a building that was designed to be its own self-contained biome. Instead of functioning as a hotel for the wealthy, the activists transform it into a homeless shelter. "The Boston Hearth Project" is the sort of hopeful speculative fiction solarpunk is known for: reimagining the functions of spaces, standing up against the twinned forces of capitalism and the state, and deploying green technologies to address social and economic injustices. It is a story of life beyond contemporary greenwashing in favor of ecologically sound technologies used for justice.

*Porvir otimista* is not just simply a sunny disposition. It also addresses a shortcoming of degrowth thinking. As its name implies, "degrowth" is antigrowth. While entirely necessary in times of climate disaster, even degrowth thinkers admit that the idea does not get a lot of traction, particularly in the world of politics. As the authors of a degrowth article published in *Nature* admit, "growth is often treated as an arbiter of political success. Few leaders dare to challenge [gross domestic product] growth."[33] As Roscam Abbing told me in an interview, whenever he talks about degrowth, people are put off. "You don't make friends with it. You need to constantly defend it. You don't inspire people with it." As much as we *need* degrowth, we don't *want* it, he told me. Growth has been associated with goodness for so long, it is difficult to imagine an alternative.

In contrast, the value of speculative, optimistic science fiction—solarpunk, or related projects, like Afro-Futurism—lies in describing the beneficial world after degrowth.[34] This genre of speculative fiction has inspired people on the fediverse, with solarpunk instances Solarpunk.moe and Sunbeam.city hosting environmentalists on the network. Sunbeam.city is notable for its members' experimentation with collective governance structures and noncapitalist funding—it has been a member of the umbrella Open Collective nonprofit. As its admin, Liaizon Wakest, told me, Sunbeam.city's vision of solarpunk is "an anticapitalist future that also integrates technology and reuse and ecology and nature." On solarpunk instances, the goal isn't technological asceticism—rather, it is a new integration of communication technologies into a post–fossil fuel future.

Solarpunk's roots in science fiction make it a future-oriented vision, a speculation about "worlds where social ecology, democratic technology, and solar, wind, and tidal energy are crucial elements for collective well-being."[35] It is also about the past. Solarpunk authors look to lost, obsoleted, or forgotten premodern, precapitalist technologies and practices to rethink current ways of life. Indigenous ways of relating to each other, to the Earth, and to energy are particular sources of inspiration. Thus, solarpunk's future involves a return to what was before—before capitalism's fantasy of infinite growth.

Solarpunks want this future now, T. X. Watson told me. This is his main argument for the value of speculative fiction. It's about what can be done in the near future, "what we can do now to build a near-term, better system," Watson told me. Solarpunks see the capacity to solve today's problems—it's shining up in the sky, it's in Indigenous ways of knowing, it's in the hope of children rejecting the destructive patterns of past generations.

## Adventures in Sunshine and Degrowth

A question remains: How? While solarpunk can help imagine a new world and inspire people to act, we need actual experiments and practices that can help bring about the sunny, postcapitalist, postinfinity world of the solarpunk. Here is the limit of speculative science fiction. Speculation is one thing; real-world implementation is another. This is one reason why degrowth thinking is valuable, because degrowth scholars are doing patient, empirical work to show the impacts of capitalism on the environment. Roscam Abbing, for example, identified the problem of the always-on quest for 9s. In doing so, he and other degrowth thinkers have identified where the damage is happening.

Once again, the people developing, running, and participating in the fediverse have an opportunity to shape noncentralized social media as an alternative to corporate social media. If the fediverse is noncentralized in terms of its governance, then can it be noncentralized in terms of its underlying use of energy? Can its members wrest control back from corporate social media and from cloud computing companies? Could a solarpunk future be implemented, at least in part, on the fediverse? And can that aid in degrowing social media?

Enter environmental and media scholar Anne Pasek. Drawing on both degrowth thinking and solarpunk optimism, she's experimenting with a new fediverse instance in her Trent University lab. She told me she's inspired by solarpunk's "optimism, DIY, spunkiness, and . . . technical grit." She values solarpunk's speculative vision. As she declared in an article written with several colleagues, "We desire a solarpunk Internet!"[36] However, she is also deeply engaged in the academic realm of degrowth analysis, publishing research papers that explore concrete ways we can reduce the growth logics of information technologies.[37] She mixes sunny hope and sober analysis: "What I am trying to do might be to reinvigorate some of the utility of thinking a little more carefully with those material constraints in making these sort of speculative cultural probes."

Pasek is a living embodiment of the solarpunk manifesto: instead of despairing at the climate disaster, she's chosen *porvir otimista.* Echoing Balkan's emphasis on Small Tech and Watson's call for near-term speculation, Pasek looks for human-scale solutions to environmental problems. She currently uses a solar-powered Raspberry Pi (a small device that is less power-hungry than most consumer devices) for her own personal data storage, and she sees this as a local, small act of coping in the Anthropocene. She described to me how she cares for her solar-powered server—brushing snow off a solar panel, installing a tiny cooling fan on the Pi—engaging in "small rituals of care I can offer up to a burning world."[38]

Similarly, her experiment with the fediverse is to construct an instance that runs on a Raspberry Pi and connect it to solar panels, and thus do for social media what she's done for her personal server: attempt to move past the fossil fuel industry and the dependence on the cloud. Can efforts like hers scale to the level of millions or billions of people sharing posts, liking, boosting? If so, what would that look like? Would it truly change our current obsession with the infinite scroll?

The challenges here are huge. For one thing, there is a glaring problem: the sun sometimes does not shine. Not only does it go behind clouds on occasion; it also goes below the horizon for hours on end.

We can find some answers to this challenge from Roscam Abbing and what he calls "degrowth-inspired heuristics." Like Pasek, Roscam Abbing also has direct experience with solar-powered websites: he helped transition *Low Tech Magazine*, an online magazine, to be completely powered by the sun.[39] *Low Tech Magazine*'s articles include examinations of the climate impact of the bicycle industry, ways to heat a home in winter, and many articles on solar panels. What's most remarkable about *Low Tech Magazine* is that it is a fully solar-powered website. Based in Barcelona, Spain, *Low Tech Magazine* currently runs on a 30-watt solar panel and a 168-watt-hour lead-acid battery.[40] The server is an Olimex computer, which is a single-board, ARM-based system with 1 GB of RAM—very similar to the Raspberry Pi computer Pasek is experimenting with.

*Low Tech Magazine* might help chart out a different fediverse. Visitors to solar.lowtechmagazine.com see a warning: "This is a solar-powered website, which means it sometimes goes offline." Even with a battery, *Low Tech* sometimes runs out of power. "We're OK with less uptime," Roscam Abbing says. Instead of the obsession with 9s and always being on, he said, *Low Tech Magazine* embraces the fact that its solar power/battery power system will sometimes be unavailable.

Pasek is thinking about social media as solar-powered. As Pasek told me, solar-powered social media would mean "living with different energy rhythms and therefore different social ones." Her experimental fediverse instance may be "up or down based on the intermittency of solar power. So, in the night, we might not be able to doomscroll!" We may not always be able to log on. This is a direct violation of the always-ons, more 9s mentality Roscam Abbing diagnosed. This might scare away some people—Pasek notes that only "weirdos" might want such a social media system. But given that the fediverse has been made by marginalized people who rethink what social media could be, there may be enough "weirdos" to rethink the always-on mentality, too. Intermittent social media that is tied to the rhythms of the sun could promote a new way of relating to our technology—diurnal scrolling, nocturnal logging off.

Another challenge Pasek identifies is storage. "In order to have a low-powered server, it will probably be a small device which will not have infinite storage on it," she said. "Folks might expect that their posts or at least their images get deleted after a certain number of days, or there might be a smaller upload cap than your average instance." Media files—images, video, and audio—are particularly storage-intensive, and if the goal is to reduce the energy requirements of social media, we must do something about them.

Once again, *Low Tech Magazine* and degrowth heuristics offer guides. Since images take up a great deal of storage space and server bandwidth, Roscam Abbing and his colleagues chose a now obsolete compression technique of dithering to reduce their size. Dithering reduces the informational content of the image, making it a smaller file size.[41] At the same time, it introduces informational noise into the image that our eyes interpret as depth and detail. The result in *Low Tech Magazine* is stylized, print magazine-style images, reminiscent of halftone printing techniques.

However, the aesthetic is not the only point: the point is about managing media files.

Pasek is also thinking about aesthetics and ways to manage media. "As someone who, once upon a time, was an art historian, I am really interested in these questions about how our aesthetic experiences and norms, particularly around technological fidelity, are part of those projects of community formation." Solarpunk's method of taking inspiration from history comes into play here. "The extent to which we could look to the past for models for the future and have that be a question of aesthetic definition," she says. Resisting the growth of media file size—think of the obsession with high-definition images and video—might mean returning to more emphasis on text as a worldmaking medium, much as people did on Web forums in the 1990s. In addition, a solar-powered fediverse would also likely mean changing people's expectations about the availability and resolution of media, Pasek argued.

Pasek's experiment is in the early stages, and it could fail. However, she told me that failure is an option. Her careful, analytical work will produce new, empirical insights into the material weight of the fediverse, exposing more areas in which the environmental impacts of these systems could be reduced.

This is where degrowth thinking, with its roots in empirical, ecologically informed economics, helps mitigate one of the shortcomings of solarpunk. Solarpunk's sunny disposition can easily be co-opted by capitalism, the very system degrowth thinkers argue is destroying the planet. Roscam Abbing alerted me to a recent example happening in Berlin: Meat Utopia, a solarpunk exhibition of artificial intelligence (AI)-generated art depicting a solarpunk future—a future sponsored by the transnational corporation Unilever.[42] This is the danger of a purely aesthetic approach to solving social problems, Roscam Abbing told me. The corporate appropriation of solarpunk is bankrupt, "like a real estate developer's vision of corporate sustainability."

While solarpunk's optimism could be easily sold back to people in the form of t-shirts and bad art, degrowth's dour antigrowth stance and insistence on empirical data can help ground speculation about the post–fossil fuel future. As Sunbeam.city's admin Liaizon Wakest told me, "One of the positives of being against something is that it is a little bit harder to rebrand." But this relationship goes both ways. The sunny speculation of the solarpunk is a good antidote from the oppositional approach degrowth often takes. As Pasek notes, we need both opposition and speculation. We need "a diversity of tactics—agonistic politics about resisting the things we don't want, and then a more generative, experimental and creative . . . building things we might enjoy but don't yet know to expect."

## The Point Is to Change It

Pasek's experiment is one example of imagining and implementing a fediverse beyond the failed vision of the infinite scroll. For his part, Roscam Abbing has been active with a fediverse server called Lurk. To reduce the server's storage footprint, the Lurk admins regularly delete media files—much as Pasek is contemplating doing with her solar-powered fediverse experiment. Likewise, Aral Balkan's Small Web project seeks to make small, personal websites that allow people to communicate via shared protocols. Think of this as implementing instances-of-one on a wide and easy-to-use scale. Sunbeam.city is working on collective governance structures to implement the political structures that degrowth thinkers call for: "forms of decision-making that are decentralized, small-scale and direct."[43]

Unlike corporate social media and the infinite scroll, the fediverse is open for experimentation. There are innovations possible with fediverse technologies that are not possible on corporate social media. This is a strong counterargument against those who look at federated, noncentralized, democratically controlled social media and say, "It's too complicated. Give me the infinite scroll!" As Wakest, the Sunbeam admin, told me, "I think a lot of the complications that are exposed in the fediverse are complications that should actually be exposed. I don't think making it easy and removing a lot of those complications is actually the direction [we should go in]." The fediverse exposes complications of social media—including its materiality—in ways that corporate social media refuses to do. Fediverse members are taking advantage of the complications to imagine, and build, a less energy-intensive future.

It is possible that solar-powered fediverse instances will help us past the climate crisis. But this cannot be assumed. Much more work needs to be done, and it will require a mix of degrowth's materialist analysis with solarpunk's speculative idealism. As solarpunk author T. X. Watson told me, we cannot indulge in either pessimism or optimism. "In order to motivate a person to take action to pursue an outcome they care about, you need them to believe it's possible for the thing to happen. So they can't be pessimists about it," said Watson. But we also "need them to believe it's possible for the thing to *not* happen. Because otherwise, why would you bother doing work? If something is inevitable, you don't need to try to make it happen." The point is to do the work of altering social media so that it does not contribute to the climate disaster.

This is a call for researchers: as both Pasek and Roscam Abbing have shown, there is a need for analysis of the environmental impact of the fediverse. It would be unwise to settle for naive triumphalism that the fediverse will inevitably be better than corporate social media when it comes to ecological impacts. Researchers should document if solar-powered instances even result in a net benefit. Researchers also should address the e-waste problem by finding ways to run social media servers on recycled and reused technologies. Or, perhaps the solarpunk/degrowth innovations of reducing media file size, slowing the rhythms of social media, and cutting out ad tech are enough to reduce the carbon impact of the fediverse, even if many fediverse servers remain hosted by cloud services. The fediverse's openness aids in our ability to research and document climate experiments.

This is also a call to artists and others who imagine different futures. As solarpunks argue, we need to believe something better is possible. Solarpunk-style speculative fiction can help fediverse members imagine and map out new, ecologically just ways to relate to information technologies. This can include other, fellow speculative fictions, such as Indigenous science fiction and Afro-Futurism.

As Sunbeam.city's admin Wakest puts it,

> Something like the fediverse, something that molecularizes the different parts as opposed to it being a multinational corporation that you're dealing with is the only path forward for making something positive. If the network is made by people and not corporations, and also the standards are interoperable . . . there can be groups of people who pave the way to using just reclaimed power and dumpstered electronics to move all their communication on that. . . . That's not possible in a centralized network. The ability to have a path forward doesn't even exist in a centralized approach. So that gives me hope in federated, decentralized spaces—even if they have the same potential for wasteful, exploitative resource use.

If degrowth thinkers and solarpunks are successful in transforming the fediverse into a greener technology, the covenantal fediverse could put an end to the infinite scroll.

Unless, that is, the infinite scroll puts an end to the covenantal fediverse. The world's largest social media company, Meta, is coming to the fediverse. I turn to that next.

# 8
# Threads

vanta rainbow black realized she had to do something. It was early June 2023. Meta—the world's biggest social media corporation, the owner of Facebook, Instagram, and WhatsApp, and worth US$500 billion—had announced its latest project: Threads, a Twitter-like microblogging service. And Meta's Threads was going to join the fediverse.

vanta wasn't surprised by this announcement. Rumors about Threads had been swirling for months. On March 10, 2023, word leaked out that Meta was building an ActivityPub-compliant, Twitter-like social media system that would allow Instagram users to post text messages.[1] At the time, details were scarce—including the name of the eventual product.[2] But by mid-year, more details emerged. A screenshot of Threads appeared online in June, showing a mockup of a Threads-to-Mastodon conversation.[3] Meta started meeting with several fediverse instance admins under a nondisclosure agreement.[4]

vanta black appeared in a previous chapter. As I mentioned then, vanta is a mutual aid practitioner, not only asking for help on the fediverse but giving it. The announcement that Meta was going to enter the fediverse prompted them to give back to the network she loves. She decided they would lead the charge to block Threads, creating the #fedipact hashtag in mid-June. Much like #fediblock and #isolateGab, this was a hashtag campaign on the fediverse, used to organize anti-Meta admins and members. vanta also built a website to collect the names of instances that pledged to preemptively and fully block Meta's Threads. As of this writing, nearly 800 admins have signed the fedipact—vanta black awards each of them a black heart emoji.[5] In addition, while they haven't signed the fedipact, dozens more instances have also publicly announced their intention to block Threads.[6]

For vanta, blocking Threads is a means to protect the covenantal fediverse, a place she feels at home. "I found the community to be such an amazing place that was so accepting and helped me come to terms with my gender and all that," they Signaled me. "I wouldn't be the person I am today without that place lol." Partially, this is due to how admins and moderators have focused on the well-being of marginalized members. The fediverse's "distributed

*Move Slowly and Build Bridges*. Robert W. Gehl, Oxford University Press. © Oxford University Press 2025.
DOI: 10.1093/9780197776711.003.0009

community moderation makes weeding out things like transphobia easier," vanta said, "and the Mastodon server covenant explicitly forbids transphobia and such so all but the most defederated of instances also do. Even if one doesn't adhere to that specific covenant, the bar was set higher from the start with the general vibe of the place. That's why I'm so worried about Threads lowering [the bar]."

vanta's concerns about Meta's poor moderation against transphobia come from personal experience. "I was on Facebook for years upon years since like the 7th grade lmao," she Signaled me. But as they struggled with mental health issues, and later a gender transition, she found Facebook users to be cruel. "After I came out, there were people sending suicide guides my way and shit lol," they told me. "And there's all sorts of pages posting transphobic memes daily, the kind of hateful stuff that gets people killed." As she writes on the About page of the fedipact, when she reported these things to Facebook's moderation system, she learned Meta would "do goddamn nothing" about them.[7] The stark contrast between Meta moderation and fediverse moderation is a major factor in vanta's call for blocking Threads.

In addition to content moderation concerns, those who sign the #fedipact or otherwise pledge to block Meta cite many other reasons for doing so.[8] vanta's fedipact site sums up several: "that time they helped facilitate a genocide. that time they helped try to rig an election. that time they did creepy behavioral experimentation on their users."[9] Others have cited news stories about how Meta abuses and exploits underpaid content moderators.[10] Still others cite Instagram's documented harms to young women's self-images.[11] Others point to Facebook and Instagram's algorithmic boosting of dangerous conspiracy theories.[12] Others point to Meta's hypocrisy in banning hate speech but selling ad space to white supremacists.[13] Still others cite Meta's practice of surveillance capitalism. For those who have signed the #fedipact, Meta has a long-standing history as a bad actor.

I happen to agree. Full disclosure: the instance I am helping run, AoIR.social, is signatory number six on the fedipact. We have earned a black heart on vanta's website.

AoIR.social is an instance for members of the Association of Internet Researchers, an academic group that has been producing cutting-edge scholarship about digital media for over two decades. When I first heard the rumors mid-2023 that Meta might join the fediverse, I polled the content moderators on AoIR.social: Should we federate with anything from Meta? The moderators—a collection of some of the world's leading Internet

Studies scholars—quickly and unanimously said "No!" I was gratified with this answer, because I strongly believe that *anything* Meta should be blocked. Based on the consensus of the moderators, when I saw the #fedipact announcement, I immediately added AoIR.social to the list. For myself and the moderators of AoIR.social, blocking Meta seems obvious.

The bulk of the people I've interviewed for this book agree. Ro, the founder of Playvicious and The Bad Space, signaled me "LOL, DON'T DO IT!" when I asked if people should federate with Threads. Jessica Rowbottom, the queer cabaret singer who runs her own instance, is a fellow blackheart—she was the fourth person to sign the fedipact. David Revoy, the creator of the *Pepper&Carrot* comic, draws pictures of Meta as the Grim Reaper, seeking to harvest the souls of fediverse members.[14] Socrates, the admin of Scholar. social, said, "between their involvement in the Cambridge Analytica scheme to subvert democracy and their implication in a literal genocide, it's hard to imagine any set of standards permissive enough that you could justify federating with them." Aurynn Shaw, the admin of Cloud Island, said, "I think [Threads] will be harmful for the fediverse in ways that a bunch of people don't yet understand." Aral Balkan of the Small Tech Foundation is downright angry about Meta adopting ActivityPub:

> Mark Zuckerberg called the people who trusted him with their data when he set up Thefacebook at Harvard "dumb fucks." His extractive business of building intimate profiles of people to exploit for profit has made Facebook a trillion-dollar corporation and Mark a billionaire. So should you trust Threads and federate with it? Sure. After all, his business always needs new "dumb fucks."

And as Del, the admin of Mastodon.art, pithily put it: "Fuck off, Zuck."[15]

Overall, blocking Meta's Threads is obvious to vanta rainbow black, a trans woman who has had terrible experiences with Facebook. It's obvious to a group of Internet Studies scholars who spent their careers studying digital media and are moderating a Mastodon instance. It's obvious to many of the most involved, dedicated members of the covenantal fediverse who have spent years trying to make the fediverse into something better than corporate social media.

But this isn't to say it's obvious across the fediverse. Meta is huge, profitable, and has a lot of skilled engineers. Because of these qualities, there is significant acceptance for Meta's Threads as a valuable part of the open fediverse. As a result, the call for a #fedipact against Threads is highlighting a tension on the fediverse: between big and small, individuals and instances, and above all between openness and the ethical covenant.

## A Triumph for Openness: The Case for Threads

Looking over the discussions that have taken place on the fediverse, on blogs, in tech reporting, and across the wider Internet, there are many arguments offered in favor of Meta joining the fediverse and against the fedipact. Fundamentally, they all center on one ideal: openness.

The admin of the Mastodon instance Fosstodon.org, Kev Quirk, notes that the anti-Meta folks are saying Meta isn't welcome on the fediverse. But, he notes, "no single person/entity owns the fediverse. So who are you to say who's welcome and who isn't?"[16] The fediverse is based on an open technical protocol that anyone can use. The fedipact signatories, Quirk argues, are violating a fundamental aspect of the fediverse: it is an open network, free to all—individuals and massive corporations alike, and each should be able to connect to the other. Echoing this, technologist Jon Gruber argues against the fedipact: "I think there are a lot of long-time Mastodon users who *like* the fact that it isn't gaining mainstream traction, and want to keep it that way." He contrasts this with the value of openness. If people want to sign the fedipact, it's fine by Gruber, "but then don't call [the fediverse] 'open.'"[17] He then amplifies his critique of the fedipact, noting its signatories are isolating themselves on "an island of misfit loser zealots."[18]

Quirk and Gruber are emphasizing that ActivityPub, the protocol that enables the fediverse to work, is an *open* social media protocol. For Meta to adopt it is an endorsement of such opennness—something Meta is not known for. Meta is known for its "walled gardens"—centralized, closed systems that can't interoperate. So when Meta starts talking about adopting open standards, people notice. Tech journalists noted it as a major shift in that company's approach. A *Verge* reporter writes, "for the largest and most centralized social service on the web, suddenly throwing open the gates to other platforms seemed like an unlikely pivot."[19] Reporters for *ZDNET* also noted the shift. "Historically, Meta has not been a public proponent of decentralized networks for its social media apps. In fact, the company's apps rely heavily on walled gardens to maximize the profit made from marketing and advertising campaigns targeted at users."[20]

Meta's walled garden approach is partially the reason Mastodon and ActivityPub even exist in the first place. Recall from a previous chapter that major corporate social media platforms (Twitter, Facebook) simply ignored the creation of ActivityPub in the years 2014–2018, writing it off as "small beans." This was despite the fact that the group creating ActivityPub invited the major corporate social media companies to participate multiple times. This absence of corporate social media involvement opened the door for smaller players to embrace open standards. When Mastodon implemented an early version of ActivityPub in 2017, it was by far the largest entity to do so, demonstrating the viability of the nascent protocol. Particularly after the purchase of Twitter by Elon Musk, Mastodon and the fediverse have demonstrated the viability of nonprofit social media based on open protocols.

Mastodon's size and popularity pale in comparison to Meta. In a matter of months, Meta's Threads grew to 100 million users, immediately making it ten times larger than the fediverse.[21] This is not to mention Meta's sheer size as a global corporation. For the creators and maintainers of ActivityPub, such a large entity adopting their open standard is quite a vote of confidence—even if it has come years after ActivityPub was first developed. As one contributor to the Social Web Incubator Community Group (SocialCG, the group now in charge of maintaining ActivityPub) declared, "It's not all days that a major—the major—organization in a market implements a new interoperability standard, in particular if the standard has been around for a few years. Meta / Threads has now begun to do that with ActivityPub and ActivityStreams. Regardless of how you may think of Meta as an organization, from a standards perspective, this is a major success."[22]

With the most powerful walled garden opening up, many on the fediverse argue that Meta's adoption of ActivityPub would bring massive amounts of new content for fediverse members. "My wife is looking forward to deleting her Instagram account once she can connect with the same folks from her Mastodon account," writes Mastodon founder Eugen Rochko. "Being able to remain in touch with over 100M people who still use Meta products out of the comfort of an ad-free, privacy-friendly platform like Mastodon is a game changer."[23] For Rochko and others, Threads's adoption of ActivityPub could mean that any fediverse member could follow a Threads user without having to sign up for any Meta accounts. This, they argue, would benefit fediverse members because they would have more content to consume from hundreds of millions of Threads users. Against those who argue for blocking Threads, one person argued, "Would you rather keep the fediverse restricted to its current population of ~8M people or scale it to ~2.35B+?"[24] With Meta on board, writes another, "a few years from now, you might be able to lookup [sic] all your friends and family on the fedi."[25]

Such openness could even work in reverse, argues the pro-Threads crowd—it could grow the numbers of people on Mastodon. How could Mastodon grow if Threads is open? The answer the openness advocates offer is this: once Threads users are exposed to federation, they will see its benefits and make a switch. As Eugen Rochko, the founder of Mastodon, put it, "The fact that large platforms are adopting ActivityPub is not only validation of the movement towards decentralized social media, but a path forward for people locked into these platforms to switch to better providers. Which in turn, puts pressure on such platforms to provide better, less exploitative services."[26] From this perspective, Threads's openness will allow the fediverse to change Meta for the better.

Fediverse developer Chris Trottier makes this argument more forcefully:

> ActivityPub poses a tangible threat to Meta's monopoly on social media. By choosing to federate, Meta might be opening Pandora's box. The moment Meta's users receive a message from a server *not* owned by Meta is the moment they're exposed to *something else* beyond Meta's control. Inevitably, this will create more diversity of ActivityPub-enabled platforms—not less. This will erode Meta's network effects. For people who use Meta, the power of decentralization—giving them more freedom—will prove revelatory.[27]

When I interviewed him, Rochko reiterated this argument. "While I'm not a fan of Meta, and I don't think anybody should be using a Meta platform . . . , I think it's a step in the right direction for them to be going into interoperability," he told me. "And if they stay true to their word, and they start federating properly with the fediverse, I would be OK with not seeing them as a competitor, but as a partner." From this perspective, Meta's entry means that their actual users are potential Mastodon members. Arguably, this is the driving impulse of Rochko, consonant with his mission to make an open-source alternative to Twitter that matches the scale of any corporate social media.

Finally, the pro-Meta advocates also argue that anyone who attempts to block Meta is ignorant of how the open Internet works. Given that Meta is, fundamentally, a surveillance capitalist corporation which makes its living selling user attention to advertisers, a key reason for fediverse instances to avoid connecting with any of its properties, including Threads, is to avoid having that company absorb the personal information of fediverse members. However, pro-Meta advocates argue that blocking Meta would do nothing to protect the fediverse's privacy. "Threads started to test ActivityPub integration this week and the fediverse is losing it's [sic] collective mind going into overdrive to block them in any way possible so they can't grab all your data," writes a software developer. "Here's the fun part: they can already do that and they definitely don't need ActivityPub to do that."[28] Indeed, Meta has been gathering data on people—including non-Meta users—for a long time by scraping the Internet and monitoring people as they visit websites.[29] As technologist Tim Chambers argues, Meta "already have a frightening amount of internal data from inside [Instagram], and . . . virtually all of the public fediverse is scrapeable, and syncing those to datasets is unstoppable—with or without defederation."[30] The argument here is that fedipact signatories are ignoring the fact that their data are out in the open—Meta will get those data whether it is blocked or not. If the fediverse is already open, why not celebrate Meta for opening up, too?

## Against Naive Openness

If Meta's adoption of ActivityPub has done anything, it has revealed a fundamental fault line of the fediverse—and who is on what side of the line. As The Bad Space admin Ro said to me, "I appreciate people showing their hand." On one side are the champions of openness. On the other is a desire to protect the covenantal fediverse.

Proclamations about the fundamental good of openness—and how those who resist is are at fault—have appeared again and again in the fediverse's history. As I showed in Chapter 3, they affected Mastodon early in its history, when it was using OStatus (short for "Open Status"), a predecessor to ActivityPub. Recall that OStatus's design reflected a vision of digital communication where openness was valued over everything else, a vision where all of us would proclaim who we are. Consequently, OStatus did not allow for private posting. This opened up marginalized Mastodon members to harassment from users of GNU social servers. In response, developers like Christine Lemmer-Webber, Amy Guy, and Evan

Prodromou included more privacy features in ActivityPub, allowing for marginalized people to address posts in ways to avoid having sensitive posts open to the rest of the network.

Arguments about openness were offered against calls for codes of conduct in the tech industry, as I showed in Chapter 4. When activists like Coraline Ada Ehmke agitated for codes of conduct to curb the tech industry's rampant harassment of women, gender minorities, and people of color, the pushback she got was that codes of conduct would destroy the tech industry's meritocracy—the idea that the tech industry is always open to anyone who can code well, no matter their identity or politics. But the anti–code of conduct argument ignored the ways in which "pushyocratic" tech development, where acrimonious and ad hominem arguments win out, actually *closed* tech spaces to marginalized people.[31] Codes of conduct signaled commitments to moderation against harassment. Mastodon's adoption of them not only for development but also for social networking sets it apart from corporate social media, which has featured poor moderation.

Arguments about openness were used (and still are used) against blocking, particularly instance blocking, the focus of Chapter 5. Blocking and defederating between instances are seen as failures of the open fediverse, reducing the ability of people to connect and engage in free speech. But as Marcia X, Ginger, and Ro demonstrated through #fediblock and The Bad Space, the call for openness merely reinforces the power of dominant groups at the expense of the marginalized. Openness leads to repressive tolerance: when all ideas are tolerated, very soon the loudest, angriest, most racist and intolerant voices will be the only ones heard.[32] Blocking bad instances—and sharing the knowledge about those bad instances—is one major mitigation against the power they would otherwise enjoy on an open network. With Mastodon and other fediverse software enabling instance blocklist importing, knowledge about the worst parts of the fediverse can be quickly shared.

As I discussed in Chapter 6, arguments about openness to a for-profit corporation—and how fediverse data are simply open to Meta to exploit—replicate arguments about the inevitability of capitalism, particularly surveillance capitalism. Those who argue Meta will get our data anyway are akin to those who argue that we might as well give our personal information to corporations, since they're going to get it anyway. The activists who have built noncapitalist social media deny this hopelessness, finding ways to fund servers and support artists without subjecting everyone involved to the gaze of corporations. Likewise, naive openness implies the same infinite boundlessness that has driven the climate disaster, the focus of Chapter 7. The call for open networks and the desire to grow, grow, grow at all costs is being denied by those who point out that there is no such thing as infinite growth.

Activists who have pushed back against naive openness have produced a new form of alternative social media: a noncentralized, covenantal fediverse. Millions of people are now freely engaging in social media activities as members of small, self-governing communities that declare their values via codes of conduct and then freely associate with other, like-minded communities. Rather than being subject to harassment from servers that don't abide by this covenant, the covenantal fediverse can simply block bad servers. The covenantal fediverse is funded through noncapitalist means, using informal and formal nonprofit organizational structures instead of selling user attention to advertisers. As a network of small communities, there is more likelihood the covenantal fediverse can find ways to engage in social media without contributing to the climate disaster. Above all, the noncentralized fediverse resists any given source of centralized power—which would include Threads.

This is the fediverse vanta is seeking to protect. "fedi has done a lot for me over the years," they texted me. "It's filled the hole in my heart where the support and comfort of a non-abusive family is supposed to go. fedipact has been the least I could do" to support it. With Threads potentially joining the fediverse, once again members of the covenantal fediverse are banding together to repudiate a bad actor. Meta has a long track record of abhorrent practices, any one of which would violate the shared ethical values of the covenantal fediverse. Indeed, as vanta told me, Mastodon.social, the flagship Mastodon instance run by Mastodon gGmBH, "would be in violation of their own server covenant if they federated with Threads."[33] The fact that Mastodon's founder Rochko himself is welcoming Threads (admittedly with some reservation) does nothing to change the minds of the fedipact signatories, since (as was pointed out by a pro-Threads advocate) no one owns the fediverse. The fedipact represents instances who are "drawing a line in the sand to make it clear we're not for sale," vanta told me.

## Individuals and Instances

Is this a situation where, on the one hand, there are the champions of openness, and thus, on the other hand, there are the fedipact signatories, who are isolated and closed off? Is this fundamentally a struggle between open and closed? Not quite. Instead, it is a conflict over a key question: What is the basic unit of the fediverse, the instance or the individual?

First, it is false to argue that the #fedipact members are against openness. As I have argued across this book, the covenantal fediverse is open to people who want to become *members.* They just need to do a few things: see themselves not as signing up for social media, but joining a community and abiding by its ethical standards. These communities are open, so long as members contribute to the community's self-governance and maybe even contribute a bit of money each month to support their local admin. Or, the fediverse is open for people to run their own instances, if that's what they would prefer. If a new instance agrees to the covenant, that instance and its members can be part of the wider network. It is an open network, but it makes demands of its members. This might sound like a lot to ask of being a member of social media, particularly after two decades of individual-centric corporate social media, but I would emphasize the *social* in social media and recognize that the covenantal fediverse is a society, and that being in a society takes effort.[34]

The covenantal fediverse is thus open, but within the bounds of an ethical covenant. This comports to what federalist political philosopher Daniel Elazar calls "federal liberty," which is "the liberty to do that which is right and proper by the terms of a particular covenant, thereby providing a basis for resolving the problem of balancing liberty and authority with which every society must grapple."[35] As Elazar argues, "every proper polity is established by a pact among its constituents which is covenantal insofar as it rests upon a shared moral sensibility and understanding and is legitimate insofar as it embodies the fundamental principles of human liberty and equality."[36] A shared covenant, Elazar argues, helps form the basic unit of federalism: the group, which establishes its norms and values and has rights to autonomy, self-determination, and to association (or, conversely, disassociation) with other groups.[37]

In the covenantal fediverse, we see covenants at the instance level in the form of codes of conduct. As I have shown, fediverse instances that establish codes of conduct use those codes for two things. First, codes set the boundaries of practice within the instances, communicating those bounds to would-be members who can choose to join or not. Second, codes of conduct shape instance decisions about associating or disassociating with other instances. As instances establish their own unique codes of conduct, connect to one another, and maintain those connections, a more global set of values emerges.[38] The emergent set of network values broadly calls for strong moderation against hate speech, disinformation, and surveillance capitalism.

This vision clashes with a more individual-centric conception of federation. In an individual-centric approach, the instance does not matter. As ActivityPub coauthor Evan Prodromou puts it, from the individualistic perspective, "The [instance] may set some parameters around content or software usage, but otherwise it's mostly a dumb pipe" that fosters "person-to-person" connections.[39] Instead of instance admins, individuals make every choice for themselves, from whom to follow to what to post. The pro-Threads group implicitly argues for individuals to make the choice of connecting to Threads users, as well as that Threads users could decide to migrate to the fediverse after experiencing federation. They argue that instance admins should have no right to determine whether instance members can make that choice. Instead of a theory of federal liberty, this approach is more of a libertarian argument, maximizing the liberty of individual members.

In contrast, the #fedipact argues that communities (in the form of instances) have the right to not associate with Threads because Meta's previous behavior strongly suggests that Threads will not comport to the ethical values of the covenantal fediverse.

## What Is Actually Being Opened Up?

However, the conflict over Threads and the #fedipact may cover up the actual objective of Meta: expanding into new markets, including the European Union. By proclaiming its support for ActivityPub, Meta can claim to European regulators that account portability—a key part of Europe's new Digital Markets Act—is implemented. Indeed, days after announcing its first implementation of ActivityPub in December 2023, Meta also announced that Threads was available in the European Union. This opened up Threads to hundreds of millions of new users.[40] By claiming to be decentralized and that it allows account portability, Meta can grow as a powerful social media center and absorb European user data.[41]

In addition to seeking new users in Europe, Meta's Threads is a play for the real customers of Meta products: advertisers, including advertisers who formerly used Twitter/X. X is hemorrhaging its advertising business.[42] As an advertising company, Meta is open for the business of former X advertisers. If

Threads can attract many users—including users in regulatory environments like the European Union—then its advertising capacities are valuable to former X advertisers.

Considering Meta's play for hundreds of millions of Europeans and hundreds of millions of advertising dollars, the real question is, Does the company care one whit about the fediverse? Is it concerned at all about a noncapitalist network of mere millions comprised of instances that average around 400 members each? Frankly, it is difficult for me to believe that Meta, with its billions of users, would concern itself at all with the fediverse, either as a network to take over or as something that could influence its decisions.[43] Instead, Meta's eyes are clearly on European regulation and winning over former Twitter/X advertisers.

If Meta doesn't care about the fediverse, what does it matter that Threads has adopted ActivityPub? Will it actively harm the fediverse? An embrace of Meta's Threads would drastically undermine the fediverse's claims to noncentralization, because Threads would instantly be the most massive part of the network and thus a powerful center. The likely outcome of the naive openness to federating with Threads is a warping of the most viable, current alternative to corporate social media, the covenantal fediverse. Those who argue Meta's Threads will provide more content for the fediverse ignore the fundamental mismatch between the fediverse and Meta's approaches. Multiple fediverse members have noted that the algorithmic approach taken by Meta could result in certain types of posts getting more engagement because they are being boosted by Meta's algorithm.[44] Meta's algorithm favors posts that attract attention—including posts that inspire anger or fear.[45] If those posts cross over from Threads to the rest of the fediverse, they will appear to have been boosted in the same manner as the fediverse's standard, reverse chronological posts, but will actually have been amplified by a Meta attention algorithm. Over time, as algorithmically boosted posts gain more attention, fediverse members will see such posts—fear and hate posts—as more important than the ones that have appeared on the fediverse prior to Meta's arrival.

In addition, Meta's use of surveillance capitalism would exploit the creativity of fediverse members—even if fediverse moderators keep ads off Mastodon and other fediverse systems. While ads from Meta might not make it to the fediverse, things can flow in reverse. If Threads users follow fediverse accounts who create posts that keep Threads users' attention on Threads, then Threads is effectively monetizing fediverse content. Meta can sell ads around content coming from the fediverse just as easily as it could sell ad space around Threads users' content. Fediverse members don't have to see ads, but their posts can be monetized all the same.

Finally, Meta could undermine a major part of ActivityPub and, in doing so, continue an unfortunate habit developed by the Mastodon project. Recall from Chapter 3 that fully one-half of the ActivityPub standard is ignored. While ActivityPub's standard for instance-to-instance federation forms the basis of the current fediverse, ActivityPub's client-to-server standard, which would enable developers to make mobile and desktop apps to access the fediverse, is not used. This is because developers adopted Mastodon's application programming interface (API), which predated ActivityPub's finalization and has been the dominant approach to mobile and desktop client development. People adopted Mastodon's API because Mastodon was a first mover and the largest player on the fediverse. This will be no more with Threads in the picture. With the use of nonstandard, non-ActivityPub APIs already established, Meta is free to dominate API access to federated social media. If a large number of fediverse instances do federate with Meta, eventually people using Mastodon API-based apps like Tusky might abandon them in favor of a Meta-developed app (which would be, if history is any guide, an app with terrible privacy policies).

In short, there are many ways Meta could deeply alter the fediverse for the worse, while there are very few benefits the fediverse would enjoy due to Meta's participation.

## Will the Fediverse Fragment?

vanta rainbow black and the fedipact have recognized the likely harms Meta will bring to the fediverse. Activists, often coming from the most marginalized populations, defend the most viable social media alternative that has ever been built from being dominated—this time by the world's most powerful social media corporation.

Like all the other struggles—for privacy for marginalized people, for codes of conduct, for the ability to block, and resisting the growth logics of digital capitalism—blocking Meta has not come easily. As vanta texted me, "I sorta figured it'd be easier to convince everyone to block Threads than it actually was. Like thinking back to Gab and all that. But even if the response has been more divided than I anticipated, the fedipact still fulfills its purpose of keeping people safe." While over 800 instances have signed the #fedipact as of this writing, they only represent 8% of all active fediverse members.[46] The fedipact is not alone, however: the sixty instances listed in the Fedi Garden also agree to block Threads, and I have seen many people discussing blocking Threads on the #fediblock hashtag.[47] However, even accounting for all these fellow travelers, instance-level Threads blocking is only happening on a minority of fediverse servers.

Meta's entry and the #fedipact highlight conflicting perspectives. One possible result is that the fediverse will fragment along the lines Evan Prodromou has identified as "Big Fedi" and "Small Fedi." Big Fedi, Prodromou argues, welcomes commercial services, sees individuals as the key social unit, and values growth. Small Fedi is more akin to the covenantal fediverse I've been describing in this book: it centers the instance, where "moderation decisions, cultural decisions, account decisions, most social decisions should happen" and eschews corporate social media.[48]

Regardless of whether one is a Big or Small partisan, Meta's Threads will need to be reckoned with because of the sheer size and resources that the corporation brings. With Threads in the picture, Big Fedi partisans now must advocate for other big corporations to adopt ActivityPub, or else see the entire ActivityPub project be dominated by Meta. Given how large Meta is, this means advocating for other Big Tech firms—such as Apple, Google, Microsoft, or Amazon—to adopt ActivityPub as a sort of counterbalance to the power of Meta. If these corporations join the fediverse, then it is hard to see how the fediverse would remain an alternative to corporate social media. Big Fedi partisans may be getting their wish, because multiple for-profit corporations are discussing or implementing ActivityPub federation. Automattic, a company that owns Tumblr and Wordpress, has started implementing ActivityPub functionality.[49] Flipboard, a company that aggregates media posts and sells ads around them, operates a Mastodon instance at flipboard.social. Medium, a company that sells subscriptions to blog posts, also operates a Mastodon instance at me.dm.[50]

For Small Fedi partisans, the covenantal fediverse can continue if it blocks Meta and other corporate social media, but at a cost: the covenantal fediverse's influence could fade as people start to see the fediverse as whatever Meta (or another large, for-profit corporation) decides it should be. If Meta or other corporate social media dominate the fediverse, then it will no longer function as an alternative to corporate social media. The covenantal fediverse might be in its final days as Meta comes to town.

This does not mean it is dead. Activists like vanta, Ro, Marcia, Aurynn, or Aral will no doubt soldier on, struggling against corporate social media by making alternatives. Fundamentally, this is because of their ethical commitments, which transcend any specific alternative social media implementation. Their commitment to caring for the fediverse is the important aspect of the covenantal fediverse. I turn to the central role of care on the covenantal fediverse next.

# Conclusion

## Caring for It: Putting the Ethics in Ethical Social Media

DJ Sundog is a musician living in rural Georgia. Sundog told me that the fediverse has helped change him. When he was growing up, he "had life on easy mode" because "I was a young white man who grew up in the church," the son of a preacher. This background made his transition to the tech sector easy, where he became a self-described "jerky techbro kid" who believed in individualistic, "pick yourself up by your bootstraps" self-reliance. Looking back on this period of his life, Sundog laughs: "I was crap, man. Like, I was not somebody who was nice to know."

When he started running his own one-person Mastodon instance in 2017 and he encountered Awoo.space—the Mastodon instance that only federates with preapproved servers—he found his libertarian senses offended: "Wait, I have to ask you permission to talk to people who are using your server and prove that my little one-person server is ok to talk to? What's that all about?" His vision of networking technology was the "great libertarian dream of everything is open and everything is free, and you will hear what I have to say, and I will hear what you have to say, because there will be no barriers between us and no censorship."

The more he learned, however, the more he realized what Awoo.space's members were doing. "That's when I found out a lot of the backstory, and of the furry community, and some of the trials that they went through on other platforms, and what they were trying to avoid with their whitelisting." As queer and trans folks, Awoo.space's members had lived lives quite different from DJ Sundog's. Through interacting with Awoo's admins, "I learned to adapt to [their] model of federating, instead of the libertarian, you-must-hear-me approach to communication." The libertarian approach, he now says, "sounds really good, but in practice, always it leads to a bad place." He notes it took him a long time to learn this, but "some people learn that earlier than others."

*Move Slowly and Build Bridges*. Robert W. Gehl, Oxford University Press. 
DOI: 10.1093/9780197776711.003.0010

He began to learn about the covenantal approach to federating, with like-minded instances-as-communities banding together through shared ethical values that they express in codes of conduct and moderation practices, where people are free to speak within the ethical bounds of their shared connections. He then recalled to me his growth as a human on the fediverse:

> I can use this to become the person I thought that I wanted to be, but never could be. . . . It's beautiful to have a space to say, "I'm going to be the best me here," and just see how that goes. . . . That's a massively powerful thing. No corporation can give you that. No critical mass of monthly active users can give you that. No marketing plan can give you that. No technology stack can give you that. It's only interacting with others, as peers, as equals, in a space that's not going to reject you for immediately for not fitting into the space they expect you to fill.

Sundog has increasingly cared about and for the fediverse, contributing code to Mastodon, maintaining a Mastodon client (Brutaldon, a minimalist text-only client that runs in computer terminals), and running an instance for music lovers called Linernotes.club. He worked with Linernotes.club members to develop a code of conduct so it would be well-federated with the rest of the covenantal fediverse.

I quote DJ Sundog at length in part because he's expressing something I heard in many other interviews—that interacting with others on the fediverse has altered their view of the world. I've heard variations on "the fedi has made me a better person" many times in my interviews.

I'm also quoting him because DJ Sundog and I are a lot alike: like him, I grew up in the rural United States, a cis, white guy who likes FOSS (I now have a beard to complete the look). Like Sundog, I'm also a bass player. I've also found my perspectives changed by my time on the fediverse—I compare my early research on alternative social media to today and see how narrow my perspectives once were.[1] I know that my own life has been wildly different from many of the people I've interviewed for this book—people who are queer, nonbinary, women, Black, trans, furry, or combinations of all of these. I also find myself in awe that in the 2020s, a space like the fediverse can not only exist, but thrive.

People like DJ Sundog and I benefit from the fediverse because many people have cared about and for it, even at great personal cost.

## The Ethics of Care

I've spent pages talking about the covenantal fediverse's shared ethical program, considering how codes of conduct shape federation connections, how moderation practices are discussed and shared among admins, and how debates over blocklists and preemptive blocking are about the tension between openness and the covenant. A question remains: What exactly is the ethical program of the fediverse? Reflecting on these practices, I believe that the covenantal fediverse relies on what moral theorists call "the ethics of care."

The ethics of care emerges out of feminist philosophy in the 1980s.[2] Perhaps the most quoted definition comes from Berenice Fisher and Joan Tronto, who argued in 1990 that care is "a species activity that includes everything we do to maintain, continue, and repair our 'world' so that we can live in it as well as possible. That world includes our bodies, ourselves, and our environment."[3] Fisher and Tronto argue that caring has been largely overlooked in Western philosophy, marginalized as a merely private act done predominantly by women and featuring only emotional competencies instead of rational ones.

Instead of seeing it as a marginal activity, Fisher and Tronto see caring as essential to human society. "Caring crosscuts the antitheses between public and private, rights and duties, love and labor," they write.[4] Care is relational, happening between caregivers and care receivers. It is thus not about a sole individual, but about people in relation to one other: we care about, care for, and take care of others. The caregiver considers and tries to meet the needs of the care recipient. And, as embodied beings who often find ourselves helpless, all of us also must receive care from others.[5] Subsequent philosophers have further developed care ethics, considering how it intersects with racial identities, how caring and masculinity might be articulated, how care is attenuated through abilities and disabilities, how caring can be found beyond interpersonal relations and in domains such as public policy, and how care also involves technical practices, such as maintaining technologies.[6]

Throughout this book, I've discussed people who care deeply about and for the fediverse, despite the potential harms. Some of them are Mastodon instance admins—Del, Aurynn, Socrates, Shork, and Mawr—who maintain instances and help people join their communities. There are producers and maintainers of the ActivityPub protocol, such as Amy Guy, Christine Lemmer-Webber, and Evan Prodromou. There are activists, including

Coraline Ada Ehmke, who challenge the tech sector's myopic focus on meritocracy and radical openness, and there are those who took up Ehmke's codes of conduct on Mastodon. There are developers and fediverse moderators challenging racism and bigotry on the network through hashtag activism and building blocklists: Ro, Oliphant, Marcia, and Ginger. There are those who use the fediverse in noncapitalist ways, including for mutual aid, as EdenDestroyer does, and for sharing art, as David Revoy and Jessica Rowbottom do. There's Eugen Rochko, who has refused venture capital for Mastodon in favor of keeping it not-for-profit, as well as the thousands of other instances that also function as nonprofits. There are those who challenge the growth logics of corporate social media, such as Roel Roscam Abbing, Aral Balkan, and Anne Pasek. There are those who push back against corporate social media, as vanta rainbow black did with the #fedipact. There are also millions of fediverse members who do little acts of care every day: adding alt text to any images they post, boosting and favoriting each other's posts, and sharing memes, art, music, cat and dog pictures, and the news of the day.

These fediverse members clearly care about, care for, take care of, and receive care from the fediverse.[7] They do these in spite of facing harms, ranging from the stresses of developing new technologies to making mistakes in (and learning from) difficult moderation decisions to receiving harassment and death threats.

Taken as a whole, the ethics of care strikes me as the fundamental ethical program of the fediverse. It contrasts with other ethical theories that valorize individuals who operate in isolation and do the universally recognized right thing after rational reflection.[8] Ethical theorists such as Immanuel Kant or John Stuart Mill put individualistic, rational men who ponder facts and then make wise, irrefutable decisions at the center of their thinking. Kantian thinking, called "deontological ethics," tends to emphasize individual rights and the duty to make an ethical choice after reflecting on the matter. For Kant, it is the individual will to do right under moral laws that is central—a position that has been influential on contemporary libertarians.[9] Mill's ethical program, utilitarianism, centers on individuals doing a sort of mathematical analysis: the act that provides the greatest happiness to the greatest number of people is the right choice. The emphasis on outcomes is called "consequentialism."

For these thinkers, care for the other is either irrelevant (Kant's focus is on the intentions of the actor, not the outcomes) or so abstract as to be meaningless (the greatest happiness for the greatest number requires math on a global scale). Kantian and libertarian thinking also put the individual's rights at the center. Care ethicists find deontological or consequentialist theories lacking because they do not consider how relationships (between people, machines, the natural world) affect ethical choices.

To be fair, deontological and consequentialist thinking can be easily found on the fediverse. Codes of conduct are arguably deontological: they spell out the moral laws of an instance (e.g., no racism, no transphobia, use alt text) and establish social protocols of behavior. Fediverse members who read codes of conducts learn their duties and responsibilities. Arguments in favor of the growth of the fediverse—including those in favor of welcoming Threads—are often couched in utilitarian terms: people will be happier if they leave corporate social media and join the fediverse, and the more people here means more content for all of us. And as is often the case with free and open-source software (FOSS) cultures, libertarian thinking is common: by running our own social media, we liberate ourselves from surveillance capitalism.

Above all, however, I agree with the observations of the Systerserver collective, a group that operates on the fediverse: the covenantal fediverse is comprised of instances-as-communities that "depend on each other, sharing their tools while fostering webs of commitment, responsibility and care."[10] The covenantal fediverse runs on care.

## The Dangers of Care

Having an ethical theory is one thing. Putting it into practice is another. As Ro, founder of Playvicious.social and The Bad Space, put it to me, "the biggest challenge [for the fediverse] is the ethical challenge." The central struggle of many of the people discussed in this book—to build a noncentralized, covenantal fediverse—continues, because caring is messy, uncertain, and never-ending. In contrast to deontological thinking ("Lying is wrong; therefore we should never lie") or the mathematics of consequentialism ("This act made more people happier; therefore it is right"), care ethics offers no simple solutions to the problems of building, maintaining, governing, moderating, and thriving on our own social media.

Care ethicists have noted that this uncertainty can be dangerous. Unlike universalized ethical theories such as deontological ethics or utilitarianism, care ethics is often localized, intimate, and without clear instructions about what is right to do. It is highly situational.[11] There are often many competing paths to caring, each with their benefits and drawbacks, and so "raising an argument about which good is best 'in general' makes little sense. Instead, care implies a negotiation about how different goods might coexist in a given, specific, local practice."[12] This means, for example, that what works on one fediverse instance may not work on another (and indeed, it may not work again in the future). There are also imbalances, not only between care givers and receivers but also over who should do the work of caring and how to recognize and reward that work. And there are the problems of people receiving care without giving back themselves, likely because caregiving is often seen as a selfless, altruistic act.[13] "Practices of care are always shot through with asymmetrical power relations: who has the power to care? Who has the power to define what counts as care and how it should be administered?"[14]

Throughout the book, there are many examples of the dangers of care: the danger of caring too much, moments when care is not recognized, and the unequal distribution of care. However, fediverse members also have many ways to negotiate these dangers.

## Caring Too Much

The danger of caring too much is that it runs the risk of undermining a central aspect of the ethics of care: relationality. Those who believe that they can solve every problem by simply caring enough run the risk of ignoring their relationship to themselves, burning out their own mental and physical health. Those who believe that the fediverse can be a neutral marketplace of ideas run the risk of tolerating the ideas of the intolerant. Those who care so much about the fediverse that they see competing visions for it as a threat run the risk of vilifying others and refusing to engage with them.

One of the dangers of caring is that someone may believe that any problem can be solved if they just cared more. The imperative of the covenantal fediverse—that it only works if everyone contributes—is the seed of this problem. Unlike corporate social media, where users sign terms of service agreements, start posting, and leave details like content moderation and hosting to the corporation, the fediverse requires people to create code, install software, write codes of conduct, and moderate—all on top of doing normal social media activities. This affords people who are dedicated to the ideals of the fediverse many opportunities to give, and of course they do.

As I have shown, caring about the fediverse can lead to burnout. Many of the people I spoke to have experienced burnout. ActivityPub coauthor Amy Guy has sworn off future standards work. Marcia X, a moderator on Playvicious.social and cocreator of the #fediblock hashtag, has repeatedly and publicly stated that they no longer will engage with the fediverse to the same degree. "I have run out of faith in social media at this point and use it like I'm an ornery old man," she told an interviewer.[15] Mastodon founder Eugen Rochko has publicly proclaimed he is burnt out. There are, no doubt, thousands of others who have quietly stepped away after making major contributions to the fediverse.

Fediverse admins and members have developed strategies to mitigate this danger. One is to share caring responsibilities among multiple people. Not only does this prevent the burden of caring to fall too much on any individual, it also comports with the noncentralization principle of the fediverse, avoiding having one person take on too much care work.[16] Instance admins often promote long-standing, dedicated members to admin status, and many instances have multiple moderators. If burnout does set in, instance admins seek out others to take over—indeed, many of the admins I interviewed were the second or third to hold that role on their instance. Finally, another approach is to shut down an instance in a responsible manner. An example here is the Mastodon instance cybre.space, which started in 2017 and was closed January 1, 2024, after over a year of warning from its admin. Rather than recruiting someone else to take it over and thus risk watching a five-year-old community be mishandled, the cybre.space admin decided to shut the instance down, giving members enough notice to migrate accounts to other instances. DJ Sundog, who knows the admin, told me, "I thought that was a brilliant decision. A painful decision. Obviously heartfelt. Obviously a lot of thought had gone into it."[17]

Another, related form of caring too much can contribute to repressive tolerance. Recall that repressive tolerance is the tolerance of all ideas, including the ideas of those who are dominant and in power.[18] In North America during a time of rising white Christian nationalism, this means tolerating the ideas of groups of people who would oppress marginalized people, such as sexual minorities, people of color, religious minorities, or Indigenous people. In addition, these ideologies call for the subjugation of women to men. Those who suggest that social media should be a marketplace of ideas, open to all—including arguments for the demise of minorities and the subservience of women—tout the dogmatic ideal of freedom of speech, where we all must listen to each other. The problem here is that this predominantly places the burden on the very groups under threat: the expectation becomes that marginalized people must show care by listening to the speech of the intolerant.

Many of the moderators, admins, and fediverse members I spoke to discussed giving others the benefit of the doubt and being willing to listen, but only to a point. By far the most work moderators and admins must do involves inquiring with other moderators and admins about posts that violate codes of conduct. Was the post purposely malicious? Or was it inadvertently harmful? If it's the latter, sometimes tense negotiations ensue over what to do. If it's the former, then what people tell me is that there is no obligation for someone to care for others who wish them dead. Jessica Rowbottom, the queer cabaret singer who runs her own Mastodon instance, put it bluntly: "I'm going to block people trying to kill me." Mawr, the admin of Plush.city, told me: "We block a lot of people. It's staggering the number of Nazi instances that pop up." This reflects the ideas of care ethicists, who argue that there needs to be some basic form of reciprocity between caregivers and care recipients—covenantal fediverse admins never expect any such reciprocity from bigots.[19]

Finally, there is another form of caring too much about the fediverse, one that continues to sap the energy of the network: when people care so much about a particular vision of the fediverse that they see those who advocate for other forms as betraying the network. Given all the ways people can shape the fediverse—adding to its free and open-source code, installing and running instances, or writing codes of conduct—it is inevitable that people will disagree about what is best. For example, the debates over shared blocklists, which date back to the earliest days of Mastodon, continue to this day. I myself support their use—they have been invaluable in running my home instance, AoIR.social—but I know people passionately argue that blocklists run the risk of putting too much power in the hands of their maintainers, citing previous examples of Twitter blocklists that harmed more people than they helped.[20] While the debate over this and other features can be civil, oftentimes it descends into flamewars. Such acrimonious flamewars can be a daily feature of life on the fediverse. In disputes over governance features, where is the line between debating ideas and destroying relations?

While I think this form of caring too much will always be present on the fediverse, I find some wisdom in noncentralization. For example, those who believe that Eugen Rochko has made poor choices in leading the Mastodon project—and there are many—have multiple outlets for their energy: they can make forks of Mastodon (such as Hometown or Glitch) and establish their own governance, or they can make different, ActivityPub-enabled microblogging systems (such as Firefish). For those who sign up for an instance and have issues with how it is moderated, they can reach out to the admins and moderators and discuss matters directly. If they still believe the instance is poorly run, they can move instances or set up their own. Those who support the use of blocklists but do not like the maintainers of existing ones can establish their own, rival lists, and of course those who do not support shared blocklist use can avoid using them altogether. Fediverse members put their care work into areas of the fediverse that they feel would benefit and appreciate their contributions.

## Inability to Recognize Care

Care ethicist Parvarti Raghuram notes that one of the major dangers of centering care is that those who do care work are not recognized for what they do. "Notions of care are imbricated in global patterns of power that go beyond 'who cares' to 'how care is defined and validated.'"[21] One person who does a specific type of care work may be praised, while another who does a different form may be overlooked. Likewise, one person who provides care work might get lauded, while another who does the exact same care work may be ignored unless they struggle for recognition.[22]

On the fediverse, the overvalorization of technical ability is often a factor.[23] For example, in her conversation with me, #fediblock cocreator Marcia X downplayed their ability to care for the fediverse—"I put so much work into [Playvicious] with the tools that I have. I don't do tech. I don't do code." #Fedipact creator vanta rainbow black said something similar. When she conceived of #fedipact, she put the idea out in a post, hoping someone with more power—an admin—could implement it. "I figured . . . I'm not an admin, so what else can I do besides release the concept into the wild and hope someone else takes it and runs with it?" they Signaled me. "I figured some instance admin would see that and realize the idea on their own." But when no admins created the pledge, vanta promoted it herself.

The anxieties of two people who helped create historically important fediverse hashtag campaigns (#fediblock and #fedipact) reveal societal bias toward technical ability or system administration as more valuable forms of intervention in digital media. The flipside of Marcia or vanta's downplaying of their own abilities is the valorization of the ideas of high-status technologists. Mastodon founder Eugen Rochko is often seen by journalists as offering the final word on Mastodon (if not the rest of the ActivityPub-based fediverse). As high-profile technology firms enter the fediverse—such as Automattic (which owns Wordpress and Tumblr), Flipboard, or Meta—their CEOs and engineers are sought out for comments and predictions. While I'm certainly not arguing that tech journalists ignore these actors, it is also clear that the vast majority of people who do care work on the fediverse are ignored because their contributions don't fit the standard we've developed over the past several decades. The perception that there are singular technologists who grace the rest of us with new technologies through their genius reflects the critique care ethicists have leveled at other ethical theories: too much value is placed on an individual acting and making the right choice, and not enough on relationships of care.

But not all technologists are recognized. Ro, the former Playvicious admin, Bad Space founder, and current head of a nonprofit called Hueristic Instruments, told me about people being surprised and confused to find out he's a skilled coder. As a Black man, he gets questions about his ability to build technologies, as well as his motives for doing so. People question his ability to provide technical care. This reflects long-standing issues Black and Indigenous people or women have faced in technical fields—the constant need to assert that, yes, I can code.

While these instances of overlooking or downgrading the value of acts of care can alienate people from the fediverse, care ethicists see opportunities in them. As Raghuram argues, "These different notions of competence are not only sites where inequalities of power come to light but also where new coalitions around care can be formed."[24] Recent scholarship has shed light on previously unheralded people who make digital technologies possible: content moderators, marginalized people who contribute to technology development, and people who make technology more accessible to those with disabilities.[25] In the FOSS world specifically, there are conversations about how to recognize the work of community managers and mentors as essential to the success of projects. As for the covenantal fediverse, since every aspect of it can be community-based, it reveals the necessity of many different forms of care work. Recognition of the care work happens every day: people thank Mastodon code maintainers, ActivityPub developers, and moderators, they donate to admins, and they boost each other to acknowledge their contributions. This is a step toward Raghuram's coalitions of care—gatherings of people who give care according to "his or her capacity for care."[26]

## Unequal Distribution

Finally, covenantal fediverse care isn't equally distributed. Even a cursory glance at the Fediverse Observer map shows that the network is simply not present in the Global South, particularly Central and South America and Africa. It is concentrated in North America, Eastern Australia, Western Europe, and South and South Eastern Asia.[27] If we do care about shifting people away from corporate social media, why not concentrate on overlooked areas?

Over the past few decades, many charitable projects have sought to provide digital technologies in the Global South. Consider, for example, One Laptop Per Child (OLPC), a charitable project started at MIT in 2005. Predicated on the idea that lack of access to digital technologies was the key impediment to development, the intention was to build low-powered laptops that would be given to children in the Global South. It is easy to imagine a similar charity starting today, arguing that if we just supplied the fediverse to the Global South, inequality would disappear. However, much like OLPC, such a project may easily ignore the actual needs on the ground in various parts of the world. Reacting to the debut of OLPC, "one attendee said she'd rather have 'clean water and real schools' than laptops."[28] Ignoring fundamental needs in favor of new technologies has been a constant trope of Western visions of addressing the problems of the Global South.[29]

For the fediverse to find a home in the Global South, it cannot simply be a matter of blithely foisting yet another technology on people. The ethics of care demands that we engage in relationship-building and do so while considering local practices and situations. There are efforts to do this sort of on-the-ground care work. For example, Gordon Gow of the University of Alberta engages in what he calls "technology stewardship" with nongovernmental organizations (NGOs) in Sri Lanka and Trinidad who are considering using social media. Using workshops to learn what the NGO members need, Gow and his students demonstrate ways in which federated social media might meet their requirements.[30] Gow and his colleagues are mindful of the danger of caring too much: he and his team do not simply push the fediverse upon groups of people who don't want it as if these would-be fellow fediverse members don't know any better. Gow's relational approach recognizes the fact that, in the end, the NGOs seeking to use social media may simply reject the fediverse in favor of corporate social media. "Launching a campaign with a community of practice to explore [a] platform like Mastodon can require considerable effort, and there is of course no guarantee it will succeed in changing existing practices in any significant way," he writes. "The training course emphasizes two other principles of technology stewardship that are essential for campaign planning: use the knowledge around you and understand failure and build on success."[31]

If the fediverse never touches regions of the world, does that mean that its members are free to not care about those regions of the world? No. "Technology has made it possible for the effects of our actions to extend far beyond the range of our personal encounters," feminist ethicist Carol Card reminds us. "We can affect drastically, even fatally, people we will never know as individuals."[32] This is as true of the fediverse as it is for many other technologies. Here, I would point to the acts of care that environmentalists, such as degrowth thinkers and solarpunks, are engaging in. Since the fediverse, like other digital technologies, can contribute to ecological destruction in the form of carbon emissions and the mining of rare earth metals necessary for digital components, environmentally conscious care work taking place at a local level can have global impacts. Switching to solar power, cutting back on growth logics, reducing or eliminating carbon emissions, and extending the life of hardware can be local acts of fediverse care that also indirectly benefit people who live in the Global South.

## The Covenantal Fediverse Might Fail

We commonly think of ethics as providing us with the rules to make the right choice in situations. If we only just gather more facts, think a little harder, and communicate more clearly, we can do right, others will recognize that we did right, and for the rest of history people will look back upon us and say, "They did the right thing." As a group of scholars write, in such "rationalist versions of the world, as in fairy tales, there tend to be happy endings.

Order, effectivity, efficiency, health or justice: in one way or another these may be achieved and if they are not, then someone is to blame."[33] Yet the world doesn't work like this. "In care versions of the world, the hope that one might live happily ever after is not endlessly fuelled. You do your best, but you are not going to live 'ever after.' Instead, at some point, sooner or later, you are bound to die."[34]

The covenantal fediverse, like all of us, is bound to die. In fact, as I stated at the start of this book, for as long as there has been corporate social media, there have been alternatives. Many of them did their best but have arguably failed, fading away as resources dwindled or they were unable to attract members. It's entirely possible the fediverse will, too. Even if it does not fade away in the short term, nothing lasts forever.

What happens when the covenantal fediverse dies?

What won't die (at least as long as humans exist) is the central act of caring—the messy, uncertain struggle of caring that helps put us in relation to one another. As James Baldwin put it, "Freedom is hard to bear. . . . It is the responsibility of free men [sic] to trust and to celebrate what is constant—birth, struggle, and death are constant, and so is love."[35] As I have argued throughout this book, those who built, maintain, continue, and repair the covenantal fediverse know this. Many of them come from the ranks of those who tried to make previous alternatives to corporate social media in the past—even if they failed before, they keep trying. Others are new to the game and will carry the lesson into the future. The covenantal fediverse might fade away, but the struggle—the care work—to build better methods of communication and community will no doubt continue.

# Epilogue

## On Godmonsters, or Looking Backward and Forward at Social Media

As I write this in March 2024, Mastodon just celebrated its eighth birthday. Eugen Rochko started it in 2016 as a hobby project, looking to build a better alternative microblogging system than GNU social, but by 2024, Mastodon has become more than that. As I've argued throughout this book, Mastodon and the rest of the fediverse have proven to be a viable alternative to corporate social media, opening up social media governance, operation, and participation to ordinary people in a form I call the covenantal fediverse.

When I talk about Mastodon and the fediverse in presentations or lectures, I'm often asked: Is Mastodon, or the fediverse more broadly, the future of social media? I'll admit that I really don't like this question, because predicting the future is hard. Being a critical academic, I often consider all the problems Mastodon and the fediverse will face, the forces that could tear it apart. There's a lot to be concerned about.

But tonight, I'm celebrating Mastodon's eighth birthday by setting aside my worries to engage in what CloudIsland.nz admin Aurynn Shaw calls "whimsy."[1] I'm having fun, watching a terrible movie along with a bunch of strangers around the fediverse.

### #Monsterdon

#Monsterdon is a fediverse hashtag used to organize a weekly collective viewing of a monster movie, such as *The Giant Claw* (1957) or *Octaman* (1971).[2] Participants follow the #Monsterdon hashtag, vote on a movie using Mastodon's poll feature, find a copy on a streaming service available in their area, and press play at a coordinated time (which happens to be Sunday night where I'm located). Then, they mock the movie—it's similar to what *Mystery Science Theater 3000* and *Rifftrax* comedians do.

*Move Slowly and Build Bridges*. Robert W. Gehl, Oxford University Press. © Oxford University Press 2025.
DOI: 10.1093/9780197776711.003.0011

This week's movie is *Godmonster of Indian Flats* (1973). It's an objectively bad movie. Set in a small town in Nevada, the film opens with a drunken sheep rancher who lays down in a sheep pen to pass out. He awakens next to a throbbing sheep embryo, the product of cross fertilization and a yellow gas from a local mine. That embryo is declared a scientific marvel by a local scientist who whisks it off to his secret lab. There, it grows into the Godmonster, a shambling, bipedal, mutated sheep. The Godmonster breaks out of the lab during a fight between a mob of white vigilantes who are hunting a Black man, Mr. Barnstable, who was framed for the attempted murder of a person and the staged murder of a dog. Later, the Godmonster is captured by that same mob. The mayor of the town, Mr. Silverdale, proposes to put the Godmonster on display to attract tourists, but the townspeople reject that idea and kill the monster.

My plot summary doesn't do justice to the sheer chaos of the film. As one critic notes, *Godmonster* "is the strangest, most surreal exploitation movie ever made. It offers up a Six Gun Territory theme park as a township without batting an eye, has its characters dressed like rejects from Disney World's Diamond Horseshoe Review and infiltrates the insanity with an eight foot, carpet covered sinister sheep that enjoys moonlit dances with the neighborhood Earth mother."[3] It is a deeply incoherent movie. *Godmonster* is really a fragmented collection of multiple, barely connected stories: one is about racism in the American West in the 1970s, focusing on how the town leaders frame Mr. Barnstable to prevent him from buying real estate there. Another is about wholesome American small-town life—and the use of high-tech surveillance and vigilante justice meant to keep it that way. "Violence in the name of justice controls the masses!" Mayor Silverdale yells at a town meeting. Yet another is about a small town being sold out by politicians—the very same ones who rejected Barnstable's attempt to buy land. Another is about the after effects of pollution from heavy industry. The mines are the source of the toxic gasses that mutate a sheep into the Godmonster. Yet another features the Godmonster, which shuffles around town, inexplicably growls like a panther, steals fruit from children, and kills by startling people who subsequently fall off cliffs. The filmmaker, Frederic Hobbs, attempts to make these disparate stories converge with a frenetic final few minutes, where the inhabitants of the small town riot and the monster is destroyed in an explosion.

The film is trying to say something, but it is incoherent in doing so. During the course of our collective #Monsterdon watch, my fellow Monsterdonners and I try to make sense of the madness through jokes and riffs. For me, the funniest (and most disturbing) part of the movie is how quickly the town turned on Barnstable after he was framed for killing a dog. The dog—which wasn't actually killed, but only played dead—was given a full and touching funeral. In response to this, I shared this with my fellow Monsterdonners: "The plan to discredit Barnstable was to get him drunk, let him shoot off a six-piece, and hope he shoots near the dog—but not hitting the dog—so the dog—which is well-trained to play dead—will make Barnstable think he killed the dog, and thus make it so no one will sell him land? Got it."

As we Monsterdonners joke, we desperately try to find a theme to *Godmonster.* If there is any theme to the movie, it is what cultural and political commentators call "toxic nostalgia"—a conservative desire to hold on to an idealized version of the past, typically one in which social hierarchies are firmly entrenched.[4] The small town the film is set in is a tourist trap, with townspeople reenacting Old West gun battles and maintaining brothels and saloons. The mayor of the town rejects a bid to purchase tracts of real estate around the town to preserve the town's Old West appearance. It is telling that when a Black man from out East attempts to buy land there, the town turns on him quickly, sending out a mob on horseback—indeed, the mob is so committed to recreating Old West–style justice that they chase Mr. Barnstable's car with their horses, rather than getting into cars themselves. The Godmonster itself is a representation of the horrors of the past the town is selectively forgetting: it is produced in part through toxins released from the defunct mines on the edge of town. This theme—the critique of toxic nostalgia—resonates with us Monsterdonners.

## Fragmentation and Nostalgia: The Future and Past of Social Media

A common observation about social media is that it is now fragmented, with small groups of people retreating to their own preferred platforms instead of gathering on one large one.[5] This line of thinking was particularly prompted by Musk's purchase of Twitter. Writing in *The Atlantic*, Helen Lewis argues "Elon Musk's actions at Twitter have irreparably fractured the service. We are now living in the post-Twitter era, literally and metaphorically." This accounts for some of the anxieties I hear about when I talk about Mastodon and the fediverse: Is it the future? Can it replicate the Twitter of old?

When people face an uncertain future, they often long for the past. In the case of social media, there is nostalgia for a time when influential people were on Twitter, and when Twitter was fun. Some of the people who have embraced Meta's Twitter-like service Threads have noted it feels like the Twitter of old. Kari Paul writes in *The Guardian* that "with Twitter getting clunkier and progressively less usable since Musk took it over, opening an app and actually being able to see and engage with content smoothly felt like a breath of fresh air. . . . I actually enjoyed using Threads."[6] Marketer Jay Clouse writes, "For the last several days, I've actually published more on Threads than anywhere else. And not because I'm chasing an opportunity—because it was more fun for me!"[7] Writing in *Medium,* Winston Ford says, "Spending an hour on Threads is like a time machine to 2008. It's your friends cracking corny jokes, begging for follows, and asking 'is this thing on?' Sure the brands and celebrities are there (a great move by Zuck to make sure your feed is populated), but for now it seems . . . dare I say . . . wholesome, compared to the alternative."[8] At least some people are excited that the world's largest social media company is stepping into the void that Twitter left when it was rebranded X.

Similar nostalgic assessments can be found in commentary on Bluesky, another Twitter-like microblogging system that was initially started by Twitter cofounder Jack Dorsey in 2021.[9] Particularly after Musk bought Twitter, many people have turned to Bluesky to recapture the old spirit of Twitter. Political science professor David Karpf argues that "Bluesky feels like a party that I don't know how I got invited to, but I see a bunch of people I know and they seem happy to see me."[10] Blogger Faine Greenwood writes that on Bluesky, "suddenly, social media felt fun again, rich with potential—the chance to make new friends, the chance to amuse people."[11] Kate Knibbs declares in *Wired* that "the most impressive feat Bluesky has managed isn't technical, but cultural. It has recreated an older, better era of the internet, one that's *actually fun.*"[12]

The longing for a single, central source of fun—something that recalls Twitter's heyday, before the current fragmentation—is nostalgic. But for many on the fediverse, it's a toxic form of nostalgia. As this book has shown, Mastodon in particular grew in relation to the faults of Twitter pre-Musk—the Twitter of #Gamergate and online harassment and trolling. Moreover, as this book has shown, many on the fediverse reject any return to domains created by corporate social media leaders, such as Zuckerberg or Dorsey—even if the new domains appear to recapture the fun of old Twitter.

The covenantal fediverse is comfortable with fragmentation and resistant to returning to an idealized social media past. Unlike the movie *Godmonster,* which both implodes under the weight of its incoherent plotlines, the fediverse is designed to avoid neat coherence. If the future of social media is fragmented, then the fediverse is well positioned. The ActivityPub technical protocol has proven to be very flexible, allowing for many different types of social media practices to connect: microblogging, picture and video sharing, and Reddit-style discussion forums, to name a few. Podcasting and gaming applications are in active development.[13] If "fragmentation" means that groups of people can join small servers, govern themselves, share various forms of media, and socialize with people across a vast heterogeneous network, then the fediverse is a successful, contemporary social media practice. Rather than lamenting fragmentation, the fediverse revels in noncentralization.

In spite of fragmentation, fun is still possible. While all these different types of applications are appearing across the fediverse, the shared underlying technical, social, and political protocols mean that people can still gather from across various services to engage in collective fun. #Monsterdon illustrates this well. From the perspective of my instance, AoIR.social, I saw #monsterdon jokes coming from forty-five other fediverse instances during the *Godmonster* viewing. These instances ranged in size from instances-of-one to small instances limited to friends to some of the largest Mastodon instances, such as Mastodon.social, with 250,000 active members. Topically, they range from instances dedicated to fandom and nerd culture to instances for artists, from instances for cybersecurity professionals to instances for academics, from instances for furries to instances for family members. Geographically, they ranged from localized (one instance is meant for Chicagoans) to being open to anyone in the world.

Most of these instances have clear codes of conduct. Some of them are run by admins who have different approaches to federation than I do—some are more restricted in who they connect to, while others (Mastodon.social being the prime example) will federate widely. Regardless, we share a set of ethical values. It wasn't just a terrible movie that brought us together, but the work of admins, moderators, and active members who crafted those codes of conduct and made connections with one another. While *Godmonster* flies apart at the seams in its frenetic ending, we #Monsterdon participants all converge from our distinct locations on the fediverse—and collectively, we try to make sense of the surreal monster movie through silly jokes. And then, when the movie is over, we can return to our regular social media lives.

While much of the commentary about Threads and Bluesky is about recovering the lost and idealized heyday of Twitter, the fediverse has soundly rejected that idea. Much as the Monsterdonners enjoyed the implicit critique of toxic nostalgia in *Godmonster*, we also reject toxic nostalgia in social media. We're having fun, but we're doing it in new ways. There is little desire to go back to corporate social media.

But this doesn't mean the fediverse will continue as a noncentralized alternative to corporate social media. Here I put my critical hat back on. The adoption of ActivityPub by Meta's Threads may mean that the fediverse will be swamped by over 100 million corporate social media users. Or, other problems could arise: perhaps disinformation during events such as the 2024 US presidential election might seep into the network and catch fediverse moderators off guard.[14] Or, as states begin to regulate corporate social media, new laws could be too burdensome for small, voluntary social media to comply with—a situation that affected Switter, a sex-worker-oriented Mastodon instance, in 2022.[15] The future of social media might not be federated—it could be centralized and corporate, just as it largely has been for the past two decades.

I might become nostalgic for Mastodon and the covenantal fediverse's heyday—for a time when people around the world established our own social media as a viable challenge to corporate social media, banding together with shared ethical values.

But at least for now, every Sunday, I have a way to escape my worries: watching terrible monster movies with my friends on the fediverse.

# Appendix

## Research Note

While corporate social media scholarship and criticism have been popular for the past two decades, scholarship and criticism of what I call "alternative social media" are somewhat rare, and rarer still is work focusing on the fediverse.[1] I'm including this methodological note so that future fediverse researchers might learn from my approach.

The approach I took to studying the fediverse was similar to the approach I took to studying the dark web in a previous book.[2] I relied on a mixture of participant observation, interviews, archival work, and software studies. Much of the work is thus anticipated by previous Internet studies scholarship. However, I will note two key areas in which fediverse research poses specific ethical challenges: the particularities of consent to be part of research on the fediverse, and how researchers should contribute funds to the fediverse if they seek to produce knowledge out of the labor of fediverse members.

### Consent

Consent is fundamental to fediverse research. As I've discussed in this volume, nonconsensual, experimental research has been conducted on corporate social media users—both by academics and by the corporations themselves—for years. Based on my observation and interviews, it is clear to me that fediverse members are very aware of this nonconsensual research, and they consider it to be unethical. Fediverse instances are starting to include prohibitions against nonconsensual research in their codes of conduct.[3]

In addition, I've observed that many fediverse members are leery of the journalist practice of quoting Twitter/X posts in articles without the consent of the author of the post. According to two media studies scholars, journalists have treated Twitter/X posts "more like *content*, an interchangeable building block of news, than like *sources*, whose ideas and messages must be subject to scrutiny and verification. Sources are interrogated; content, on the other hand, is simply reproduced."[4] The reduction of a post to "content" not only means that posts aren't scrutinized; it also means journalists aren't considering how taking someone's post out of one context (a discussion on a microblog) and plopping it into another (e.g., a story about how "the Internet reacts" to an event) is a violation of consent.[5]

Observing fediverse members' disdain for unethical research and the contentification of social media posts, I sought consent from fediverse members. In the case of interviews, I engaged in standard institutional review procedures for informed consent and drew on Internet research ethics guides.[6] My research plans were submitted to institutional review boards at the three universities I've worked for over the past four years: Utah, Louisiana Tech, and York. In all cases, the project was deemed exempt from full review because

I created an informed consent form and an interview protocol that stressed the voluntary nature of interviews. All of my interviewees were supplied with informed consent forms that described in straightforward terms how I would conduct interviews, and during the early part of each interview I stressed the voluntary nature of the conversation.

Part of this process involves learning how to refer to interviewees. All interviews began with a conversation about names and preferred pronouns. Some interviewees chose to use legal names, other screen names, and all interviewees shared their pronouns. This is also a key part of the consent process.

As for the issue of contentification, while I certainly gathered a great deal of data during my participant observation (journals, notes, and screenshots/captures of fediverse posts), I never directly quoted any fediverse post without the explicit consent of its author. I recommend others do the same. I realize many will protest that fediverse posts are public, but I do not think that the folks I interviewed would have agreed to be interviewed if I had the reputation of someone who takes fediverse posts out of context without consent. I did make some rare exceptions to this rule, however: if a post was already widely quoted in news coverage (as in the case of Eugen Rochko's 2018 "there's the door" post quoted in Chapter 1), I treated that as public.

Finally, I stored the interview data and notes on a self-hosted Nextcloud server, and I transcribed interviews myself. I did this to avoid storing data or passing interview audio through third-party cloud services, because I am certain few if any fediverse members would consent to have their ideas supplied to train large language models (LLMs), as is now happening to data stored in Google or Microsoft services. I realize that this book will be digitally scanned, and thus artificial intelligence companies may be tempted to use it in their training data. For what it's worth: *I do not consent to have any portion of this book be used to train an LLM or similar system.*

## Subscribing

There is a significant batch of material I've drawn on that is not readily accessible unless one pays for it. I have had the privilege of generous research funding in the past few years. Such funding allows me to buy books and computers. I also believe the funding should be used to purchase art, writings, and music from fediverse members. To that end, I have subscribed to a range of fediverse members' Patreon or Kofi accounts to gain access to these materials, paying them each month.

This obviously benefits me because I get insights into how they view the fediverse, among many other topics. Studying the creative work of fediverse members helps me develop interview questions. But it also benefits the fediverse members with some funding, which is often scarce in a noncapitalist network. I would suggest other fediverse researchers share research funds with fediverse members where possible. I would especially recommend that researchers who access large fediverse datasets via application programming interfaces or other tools also distribute research funding to fediverse members, particularly instance admins, since admins are paying the bills for bandwidth. As I write in this book, much of this network refuses the political economy of surveillance capitalism in favor of noncapitalist and mutual aid economies. Academics who derive knowledge based on the fediverse have an ethical duty to contribute funding to this network.

There is an ethical issue that might arise due to my providing funds to fediverse members who are also potential interviewees: Are the people I subscribe to obliged to talk to me because I am paying them? When I requested interviews with people I was paying with subscription fees, I made it crystal clear that my subscribing to an artist or writer does not oblige them to talk to me, and that if I reach out with an interview request, their response will not result in my cancelling a subscription. I committed to keeping my subscriptions going for at least a year, regardless of response to interview requests. (The only reason I would have ceased is if I lost my funding, but fortunately I did not.)

# Notes

## Introduction

1. My doctoral dissertation was a critique of corporate social media; it was revised into my first book, Gehl, *Reverse Engineering Social Media: Software, Culture, and Political Economy in New Media Capitalism.* I concluded *Reverse Engineering* with a call for "socialized media," highlighting alternative social media projects such as diaspora* and Lorea.
2. I'm using "ordinary" in the sense Nick Couldry uses it, where there has been a divide between media celebrities and non-media participants. Social media promises to change that dynamic, but one part of the equation is missing: ownership. We can become the media in social media, to use Couldry's phrase, but we cannot own social media. Except, however, when it comes to the alternative social media, as I will show in this book. See Couldry, *Media Rituals: A Critical Approach.*
3. Gehl, "Alternative Social Media: From Critique to Code."
4. Gehl and Snyder-Yuly, "The Need for Social Media Alternatives"; Gehl, "The Case for Alternative Social Media."
5. Atton, *Alternative Media*; Rodríguez et al., "Four Challenges in the Field of Alternative, Radical and Citizens' Media Research"; Couldry and Curran, *Contesting Media Power*; Ratto and Boler, *DIY Citizenship*; Renzi, *Hacked Transmissions*; Holt, *Right-Wing Alternative Media.*
6. Rodríguez, "Studying Media at the Margins: Learning from the Field."
7. Rodríguez, "Citizens' Media," 2.
8. Gehl, "The Case for Alternative Social Media."
9. E.g., Gorwa, "What Is Platform Governance?"; Schneider, *Governable Spaces*; Waldman and Martin, "Governing Algorithmic Decisions"; Duguay et al., "Queer Women's Experiences of Patchwork Platform Governance on Tinder, Instagram, and Vine," April 1, 2020.
10. Zuckerman, "How Social Media Could Teach Us to Be Better Citizens," 40.
11. Gehl, *Reverse Engineering Social Media: Software, Culture, and Political Economy in New Media Capitalism.*
12. I was especially inspired by Dyer-Witheford, *Cyber-Marx: Cycles and Circuits of Struggle in High-Technology Capitalism*; Terranova, "Free Labor: Producing Culture for the Digital Economy"; Fuchs, "A Contribution to the Critique of the Political Economy of Transnational Informational Capitalism."
13. Bucher, *If... Then*; Massanari, *Gaming Democracy*; Zuboff, *The Age of Surveillance Capitalism.*
14. Gehl, "Critical Reverse Engineering: The Case of Twitter and TalkOpen," 148.
15. Gehl, "The Case for Alternative Social Media"; Gehl, "Critical Reverse Engineering: The Case of Twitter and TalkOpen"; Gehl, "Building a Better Twitter"; Zulli et al., "Rethinking the 'Social' in 'Social Media.'"
16. Gehl, "Alternative Social Media: From Critique to Code," 330.
17. There have been several projects attempting to poll the population of the fediverse, which is a challenging task because it is not centralized. See https://fedidb.org/, https://fediverse.observer/stats, https://fediverse.party/en/fediverse/, and https://the-federation.info/. I will draw predominantly on the first two for estimated numbers of servers and accounts. On March 10, 2024, Fedidb counted 27,101 instances with 9.8 million accounts. See https://fedidb.org/. Fediverse Observer counted 21,765 servers with 13.9 million accounts. This may be a conservative estimate, however: Mastodon User Count reports there are 15 million accounts on Mastodon alone. See https://mastodon.social/@mastodonusercount.
18. Gehl, "Building a Better Twitter."
19. The archive is still online at https://www.socialmediaalternatives.org.
20. Gehl, "Building a Better Twitter"; Gehl, "Critical Reverse Engineering: The Case of Twitter and TalkOpen"; Gehl, "Power/Freedom on the Dark Web: A Digital Ethnography of the Dark Web Social Network"; Gehl, *Weaving the Dark Web: Legitimacy on Freenet, Tor, and I2P.*

21. Of course, most people using email today are on Google or Microsoft, with one research team recently reporting that those two accounted for nearly 40% of email accounts worldwide. See Patringenaru, "Who's Got Your Mail?" However, there are many other providers available, including domain provider-hosted emails (predominantly run by GoDaddy) and providers in specific regions, such as Tencent in China and Yandex in Russia. Self-hosting email is still possible, although those who do it report that it is difficult because of spam filtering. The problem of concentration in communications systems that run on shared protocols will obviously be one of the concerns in this book.
22. As Cory Doctorow has famously argued, there is a subsequent phase corporate social media subject people to after they are locked in: "enshittification," where those who are locked in face reduced quality service and increasing prices. As he writes, "If [corporate platforms] know we can't leave for a competitor, if they know we can't sue them, if they know that a tech rival can't give us a tool to get our data out of their silos, then the expected cost of mistreating us goes down. That makes it economically rational to seek out ever-more trivial sources of income that impose ever-more miserable conditions on us. When we can't leave without paying a very steep price, there's practically a fiduciary duty to find ways to upcharge, downgrade, scam, screw and enshittify us, right up to the point where we're so pissed that we quit." See Doctorow, "Pluralistic."
23. See note 17 for sources on fediverse population.
24. Monea, *The Digital Closet*, 116.
25. See note 17 for sources on fediverse population.
26. Lemmer-Webber et al., "ActivityPub."
27. The distinction between noncentralization and decentralization comes from federal political philosophy. In federalist political theory, the term "noncentralization" is used to refer to how "power is diffused. Centralizing or concentrating power in a federalist system risk breaking the structure and spirit of the constitution. It negates the whole telos of federalism." See Moots, "The Covenant Tradition of Federalism."
28. Reddit is instructive here. In the first year or so of my study of Mastodon and the fediverse, whenever I would talk about instances and federation, people would often say, "Oh, that sounds like Reddit." Yes and no.

    Reddit is similar to the fediverse in that it contains smaller suborganizations, called subreddits. As J. Nathan Matias (2019) documents, subreddits have a great deal of autonomy in setting their local moderation standards, recruiting moderators, and enforcing those standards. This relatively autonomous, local structure has led attorney Vincent Marrazzo (2023) to declare Reddit to be an example of federalism, much in the sense I am doing here: "Reddit's unique structure goes beyond mere decentralization and is, in fact, a governance model that is analogous to American federalism." Marrazzo's argument is that Reddit—which is a centralized corporation—operates much as the US federal government does, providing a sort of constitutional order in which decentralized subreddits can operate with some autonomy (much as US states do).

    However, as Matias (2019) and Potter (2024) have documented, and has been made increasingly clear due to recent events, Reddit's central control is far more potent than Marrazzo implies. Over the past decade, Reddit's corporate center has asserted strong control over subreddits, either to enforce content moderation policies or, more recently, in reaction to protests over changes to Reddit's application programming interface (API) and to sell Reddit data to artificial intelligence (AI) companies. Thus, in the case of Reddit, the term *decentralization* is apt: Reddit might decentralize moderation practices to subreddits, but the center can reassert itself at any point.

    Mastodon and the fediverse, by contrast, are *noncentralized* in that there are few to no mechanisms by which a central authority could assert control over the entire network. Instead, the network comprises small servers that connect to one another via the ActivityPub protocol and through social protocols of an emerging ethical covenant. Covenantal federalism as theorized by Daniel Elazar does not contemplate a strong central government so much as a banding together of small political units through ethical alignment—it is a more emergent structure than a top-down one.

    See Potter, "Bad Actors Never Sleep"; Marrazzo, "The Federalists of the Internet? What Online Platforms Can Learn from Reddit's Decentralized Content Moderation Scheme"; Matias, "The Civic Labor of Volunteer Moderators Online."
29. Zulli et al., "Rethinking the 'Social' in 'Social Media.'"
30. On March 10, 2024, Fedidb counted 11,297 Mastodon servers. See https://fedidb.org/software/mastodon. That same day, Fediverse Observer counted 10,600. See https://fediverse.observer/stats. Note as well that as of this writing (early 2024), if you search "Mastodon active users" in

Google, you will see a dozen or so articles claiming users are now "fleeing" Mastodon. I address this misconception in a later chapter. Here I will note that, while Mastodon's total member base is down since the post-Musk peak, it is dramatically elevated above where it was in mid-2022. Even if Mastodon's total is off the peak, it is not evidence that people are fleeing so much as a percentage of people have tried it and decided to not continue. It is also possible some of those Mastodon members have switched to other ActivityPub-compatible systems, such as Pixelfed or Misskey.

31. Especially Baym, *Tune In, Log On*; Barratt and Maddox, "Active Engagement with Stigmatised Communities Through Digital Ethnography"; Coleman, *Hacker, Hoaxer, Whistleblower, Spy*.
32. Gehl, *Weaving the Dark Web: Legitimacy on Freenet, Tor, and I2P*, 12.
33. Fuller, *Software Studies: A Lexicon*, 5.
34. I drew heavily from the approach outlined in Rubin and Rubin, *Qualitative Interviewing*.
35. E.g., Light et al., "The Walkthrough Method"; Duguay et al., "Queer Women's Experiences of Patchwork Platform Governance on Tinder, Instagram, and Vine," June 19, 2018; Bucher, "Want to Be on the Top?"; Bivens, "Under the Hood."
36. In a sense, I have replicated the approach taken by ethnographer Tom Boellstorff. See Boellstorff, *Coming of Age in Second Life: An Anthropologist Explores the Virtually Human*.
37. Dame-Griff, *The Two Revolutions*, 208.
38. Gehl, "Building a Better Twitter."
39. Elazar, "Covenant"; Moots, "The Covenant Tradition of Federalism"; Horowitz, "Daniel J. Elazar and the Covenant Tradition in Politics." For a discussion of how this theory applies to the fediverse, see Gehl and Zulli, "The Digital Covenant."
40. Elazar, "Covenant," 218. Note that Elazar's quote concludes that these oaths are "witnessed by the relevant transcendental authority," e.g., a deity. This is not the case on the covenantal fediverse, which relies instead on an emergent set of ethical values, rather than an appeal to a transcendent authority.
41. Mansoux and Abbing, "Seven Theses on the Fediverse and the Becoming of FLOSS," 134.
42. Fediverse developer Darius Kazemi has called for this structure. In his 2019 essay "Run Your Own Social," he says, "I would like to see groups of servers that band together through a kind of mutual approval system." His specific vision has not come to pass—he is calling for a structure he calls "neighborhoods," or groups of instances that can share posts only within that network. However, as I will show, his call for mutual approval among servers has been largely implemented via codes of conduct and instance blocking. See Kazemi, "How to Run a Small Social Network Site for Your Friends."
43. This echoes a point Helder de Schutter makes about federations of states: "Citizens in federal states are simultaneously citizens of two peoples who each exercise sovereignty: a federal people and a sub-state people." Federal political theory is a way to recognize this dual identity while still holding a larger system of states (or, for my purposes, instances) together. See De Schutter, "Federalism as Fairness," 2.
44. Marcuse, "Repressive Tolerance."
45. Mansoux and Abbing, "Seven Theses on the Fediverse and the Becoming of FLOSS," 130.
46. There is a danger in investing too much power in admins of instances, however. Nathan Schneider sees the "sysadmin" as akin to an implicit feudal lord. At least on Mastodon, the danger of the all-powerful admin is mitigated somewhat by account portability, as well as the use of external tools to govern instances. Schneider, *Governable Spaces*, 100.
47. Dame-Griff, *The Two Revolutions*, 18.
48. Flowers, The Whiteness of Mastodon; Captain, "Can Mastodon Be a Twitter Refuge for Marginalized Groups?"; Valens, "Mastodon Is Crumbling—And Many Blame Its Creator."
49. McIlwain, *Black Software*; Brock, *Distributed Blackness*; Steele, *Digital Black Feminism*.

## Chapter 1

1. E.g., Peters, "The Hunt for the Next Twitter"; Perez, "Mastodon's Latest Release Makes the Open Source Twitter Alternative Easier to Use."
2. All interviews for this book, whether conducted over video chat, phone, email, Mastodon direct message, or Signal message, are done after the interviewees reviewed and signed an informed consent form. Quotes are used with permission of the interviewees, and I use people's preferred pronouns. I also have undergone institutional ethics review for this project at three universities (Utah, Louisiana Tech, and York). In addition, all fediverse posts quoted are used with the explicit permission of their authors, with the exception of some specific posts which have already been widely quoted in press coverage of Mastodon. See the Appendix for more about the importance of consent on the fediverse.

3. Gehl, "Critical Reverse Engineering: The Case of Twitter and TalkOpen"; Gehl, *Reverse Engineering Social Media: Software, Culture, and Political Economy in New Media Capitalism*; Gehl, "Building a Better Twitter."
4. This is pretty typical in the world of FOSS projects, where many little projects work in a similar domain, each taking a slightly different approach. In this case, the variations involve making a microblogging site similar to Twitter. In the case of Mastodon, it was meant to be a more attractive design than the Spartan-looking GNU social. It was also different on the backend, implemented in the coding language Ruby instead of PHP.
5. For more about the Free Software movement, see "What Is Free Software?"; Coleman, *Coding Freedom*.
6. The interview was conducted during research for Gehl, "Building a Better Twitter," 75.
7. Ibid., 62.
8. Ibid., 62.
9. Ortega, "GNU Social."
10. This wiki page from early 2017 counted 300 active GNU social instances: "List of Independent GNU Social Instances."
11. GNU social is still active, but it is far less so than it was between 2010 and 2016. See https://fediverse.party/en/gnusocial/ for more about it. At that time, the underlying protocol was OStatus. More on that in Chapter 3.
12. Gorur et al., "Politics by Other Means?," 2.
13. Bielenberg et al., "The Growth of Diaspora—A Decentralized Online Social Network in the Wild."
14. Freitas, "Twister—A P2P Microblogging Platform"; Prodromou, "This Is My First Post"; Bielenberg et al., "The Growth of Diaspora," 13–18; Gehl, "Building a Better Twitter."
15. NotJohnMastodon, "When You First . . ."
16. Rochko, "Show HN: A New Decentralized Microblogging Platform."
17. Another difference between GNU social and Mastodon was in implementation: GNU social was written in the coding language PHP, while Mastodon used Ruby.
18. In this way, Mastodon is relatively unique in the world of FOSS alternative technologies. As Stevenson and Valente Pinto (2024) argue, FOSS alternative projects often avoid smooth interfaces, seeing them as too tied to corporate software. This sort of aesthetic distinction is one way the alternatives set themselves apart. GNU social certainly illustrates this. In contrast, Rochko based Mastodon's interface on TweetDeck, a popular, proprietary social media dashboard. Many of the longtime Mastodon members I've spoken to have noted that the attractive user interface was a big reason they saw Mastodon as a viable project. See Stevenson and Valente Pinto, "Distinction and Alternative Tech," 9.
19. Henry, "Mastodon Is Dead in the Water."
20. Rochko, "Show HN: A New Decentralized Microblogging Platform."
21. Newton, "Mastodon.social Is an Open-Source Twitter Competitor That's Growing Like Crazy."
22. Zuboff, *The Age of Surveillance Capitalism*.
23. Gehl, "Alternative Social Media: From Critique to Code."
24. Poo-Caamaño et al., "Herding Cats in a FOSS Ecosystem."
25. In this article, coauthors and I refer to this as a problem of "abstraction": Zulli et al., "Rethinking the 'Social' in 'Social Media.'"
26. In a *Medium* post titled "Learning from Twitter's Mistakes," Rochko shared safety and moderation ideas he had been provided by Mastodon members. See Rochko, "Learning from Twitter's Mistakes."
27. Chess and Shaw, "A Conspiracy of Fishes, or, How We Learned to Stop Worrying About #GamerGate and Embrace Hegemonic Masculinity"; Burgess and Matamoros-Fernández, "Mapping Sociocultural Controversies Across Digital Media Platforms." To be fair, Twitter wasn't the only site for coordination of harassment. Reddit was also implicated, as well as more fringe sites such as 8Chan. See Massanari, "#Gamergate and the Fappening." As I will show, the fediverse isn't immune to such practices, although activists have built safety tools, such as blocklists, to help mitigate the problem.
28. Gehl and Lawson, *Social Engineering: How Crowdmasters, Phreaks, Hackers, and Trolls Create a New Form of Manipulative Communication*; Gallagher, "Advanced Meme Warfare—July 6, 2016 (CW)"; DiResta et al., "The Tactics & Tropes of the Internet Research Agency"; Howard et al., "The IRA, Social Media and Political Polarization in the United States, 2012–2018."

29. Massanari, *Gaming Democracy*. Also see a documentary about the cartoon character Pepe the Frog: *Feels Good Man*.
30. Hampton, "The Black Feminists Who Saw the Alt-Right Threat Coming"; Gallaher, "Mainstreaming White Supremacy"; Morstatter et al., "From Alt-Right to Alt-Rechts."
31. Emerson, "As Literal Nazis Abuse Twitter, CEO Asks How to Improve the Platform."
32. Ibid.
33. Newton, "Mastodon.social Is an Open-Source Twitter Competitor That's Growing Like Crazy."
34. Ibid.
35. Rhodes, "Like Twitter but Hate the Trolls?"
36. Lekach, "The Coder Who Built Mastodon Is 24, Fiercely Independent, and Doesn't Care About Money."
37. Jeong, "Mastodon Is Like Twitter Without Nazis, So Why Are We Not Using It?"
38. Reflecting this legal restriction, Mastodon.social's precursor to what would be called its Code of Conduct specifically banned "Content illegal in Germany and/or France, such as holocaust denial or Nazi symbolism" and "Conduct promoting the ideology of National Socialism." See "Mastodon.social—Mastodon."
39. Most cite the Mastodon User Count account for the 1-million-member figure: https://lou.lt/@mastodonusercount/99099871022836220. Mastodon User Count is now located at https://mastodon.social/@mastodonusercount. As for historical data on the number of instances, that's a bit trickier to cite. In 2017, I collected snapshots of The Kinrar's Instances.social site, which listed at least fifty distinct instances by the end of that year. Several students and I also collected documents from Mastodon instances in mid-2017, finding around one hundred distinct instances. See https://www.socialmediaalternatives.org/archive for more. Allie Hart, a fediverse developer, reported over one thousand in her essay "Mourning Mastodon." See Hart, "Mourning Mastodon."
40. Tilley, "One Mammoth of a Job."
41. Rochko, "Learning from Twitter's Mistakes."
42. For example, in 2019, a wave of Indian Twitter users shifted to Mastodon due to protests over Twitter's banning of a prominent lawyer. See "Twitter in India."
43. The "benevolent dictator for life" term is a somewhat tongue-in-check way to describe having a project founder make final design decisions. When he was asked about this organizational model by a reporter from *Verge*, Rochko replied, "Honestly, I subscribe to that model. I think it's effective, and I think that it leads to a better product. A good product needs long-term vision and it needs cohesive vision. That's something that a committee cannot give." See Patel, "Can Mastodon Seize the Moment from Twitter?"
44. Nate Schneider sees the roots of such organizational structures in both the structure of Git-based software development, which leaves a "power vacuum" ripe for a single person to dominate, and the emphasis on sysadmins as the bosses of online services. As a Github-based online service, Mastodon certainly can fit this mold. However, Schneider's concern is more genealogical: he is tracing the concentration of power of centralized, corporate social media back to the smaller practices of Github software development and bulletin board service (BBS) sysop administration. Whether or not Mastodon *instances* fit the mold of implicit feudalism is another matter—indeed, Schneider is helping build a cooperatively run Mastodon instance called Social.coop. See Schneider, "Admins, Mods, and Benevolent Dictators for Life."
45. carrotcypher, "AMA with Eugen Rochko, Founder and Lead Developer of Mastodon, a Decentralized, Open-Source Social Media Platform Based on Open Web Protocols. Ask Your Questions Here!"
46. Tiara, "Mastodon 101."
47. Captain, "Can Mastodon Be a Twitter Refuge for Marginalized Groups?"
48. This post has been widely screenshotted and preserved. It can be seen here: Valens, "Mastodon Is Crumbling—And Many Blame Its Creator."
49. Chan, "Open-Source Developers Are Burning Out, Quitting, and Even Sabotaging Their Own Projects—and It's Putting the Entire Internet at Risk," March 19, 2022.
50. Fediverse developer Allie Hart, who was involved early on with Mastodon but later stepped back, documented a long list of requested features in early 2017, including pinned posts, alt text for images, private posts, instance blocks, and account migrations. Hart notes that these features were requested by the marginalized members of Mastodon early on. As Hart writes, "Every one of these features remains unimplemented at time of writing (and the list goes on)." This is because of the leaders of the project: "I bring up such long-standing issues because they begin to

hint at a disconnect between Mastodon's original base and those calling its shots—between its body and its head, so to speak," she writes. "Although the queer community on Mastodon made up a significant portion of its early adopters and have contributed to the project in meaningful ways, they have never held any real decision-making power." However, all of these features have since been integrated into Mastodon. I do not take this to mean Hart is being impatient; instead, I would suggest that the advocacy Hart and others have engaged in has pushed Mastodon to improve, even against the wishes of the "head" of the project, even if it takes a great deal of time and effort. See Hart, "Mourning Mastodon."

51. One person heavily involved in these processes is media studies scholar Nathan Schneider, who studies online governance. While Schneider is supportive of Mastodon as a way to leave corporate social media, he is critical of the "implicit feudalism" of its design. As he writes of the work to make a Mastodon instance amenable to collective governance, "In order to assemble a governable stack, we had to make the internet bend over backward and require our members to create way too many accounts. Even then, our self-governance has continued to feel like a necessary hack, like we are always paddling upstream rather than following a natural flow." Schneider recommends changes to fediverse instance software to improve its capacity for collective governance. See Schneider, *Governable Spaces*, 100.
52. Technologist Jon Pincus has written at length about the need for forks of Mastodon. "The official Mastodon software has a long history of *not* prioritizing safety or the needs of small-to-medium size community-focused instances," he argues, "so a hard fork has a chance to make rapid progress on all these fronts." See Pincus, "Fork It! It's Time for a Mastodon Hard Fork (UPDATED)."
53. An example is Wikipedia, which was forked early in its history, resulting in Wikipedia adopting a nonprofit model. See Tkacz, "The Politics of Forking Paths"; Gehl, *Reverse Engineering Social Media: Software, Culture, and Political Economy in New Media Capitalism.*
54. For scholarship on context collapse in social media, see Sugimoto et al., "Friend or Faculty"; Marwick and boyd, "I Tweet Honestly, I Tweet Passionately."

## Chapter 2

1. Elon Musk [@elonmusk], "The Bird Is Freed."
2. Kang, "What Elon Musk Doesn't Know About Free Speech."
3. Elon Musk [@elonmusk], "Free Speech Is Essential to a Functioning Democracy. Do You Believe Twitter Rigorously Adheres to This Principle?"
4. Martin, "Elon Musk Calls Himself a Free Speech Absolutist. What Could Twitter Look Like Under His Leadership?"
5. Chess and Shaw, "A Conspiracy of Fishes, or, How We Learned to Stop Worrying About #GamerGate and Embrace Hegemonic Masculinity"; Massanari, "#Gamergate and the Fappening."
6. Kramer et al., "Experimental Evidence of Massive-Scale Emotional Contagion Through Social Networks"; Stark, "Algorithmic Psychometrics and the Scalable Subject."
7. Briant, "Cambridge Analytica and SCL—How I Peered Inside the Propaganda Machine"; "Further Supplementary Written Evidence on Cambridge Analytica, Leave.EU and Eldon Insurance"; Cadwalladr and Graham-Harrison, "Revealed"; Gehl and Lawson, *Social Engineering: How Crowdmasters, Phreaks, Hackers, and Trolls Create a New Form of Manipulative Communication.*
8. Wylie, *Mindf*ck*; Kaiser, *Targeted*; *Revealed.*
9. This is not to say that Cambridge Analytica-style microtargeting is dead. As Emma Briant has documented, the "influence industry" is going strong among political communicators, and much of it recapitulates the psychographic targeting Cambridge Analytica claimed to engage in. See Briant, "Researching Influence Operations."
10. Katz and Shapiro, "Systems Competition and Network Effects."
11. Brinberg et al., "Screenertia"; e.g., Mantere and Raudaskoski, "The Sticky Media Device"; Zulli, "Capitalizing on the Look"; Bucher, "Want to Be on the Top?"
12. Hoover, "Twitter Users Have Caused a Mastodon Meltdown."
13. Frenkel and Conger, "Hate Speech's Rise on Twitter Is Unprecedented, Researchers Find"; Frenkel and Thompson, "Some Far-Right Accounts on Twitter Saw Surge in New Followers, Researchers Say."; Hickey et al., "Auditing Elon Musk's Impact on Hate Speech and Bots."
14. Hardcastle, "Twitter Admits 'Security Incident' Broke Circle Privacy"; "Twitter Insiders."
15. Wong, "Banning Trump Won't Fix Social Media."
16. O'Carroll, "Disinformation Networks 'Flooded' X Before EU Elections, Report Says"; Goujard, "EU Charges Elon Musk's X for Letting Disinfo Run Wild."

17. "Musk Fires Outsourced Content Moderators Who Track Abuse on Twitter."
18. Treisman, "Got a Question for Twitter's Press Team?"
19. Thorbecke, "Layoffs, Ultimatums, and an Ongoing Saga over Blue Check Marks."
20. Paul et al., "Twitter Suspends Accounts of Several Journalists Who Had Reported on Elon Musk."
21. Kerr, "Twitter Once Muzzled Russian and Chinese State Propaganda. That's Over Now"; Brodkin, "Musk Labels CBC '69% Government-Funded' as More News Outlets Quit Twitter"; Allyn, "Elon Musk Says NPR's 'State-Affiliated Media' Label Might Not Have Been Accurate."
22. Peters, "Twitter Is Blocking Links to Mastodon."
23. Gehl, "Elon Musk's Stance on Free Speech Doesn't Include Competition to Twitter."
24. Fung, "Elon Musk's Twitter Blocked Links to Rival Mastodon. That Could Raise Alarms Among Regulators"; Gerken, "Twitter Blocks Users from Sharing Mastodon Links."
25. See the Introduction, note 17 for sources on fediverse population growth.
26. Zulli et al., "Rethinking the 'Social' in 'Social Media.'"
27. Hoover, "The Mastodon Bump Is Now a Slump"; Marcelline, "Mastodon Sees Dip in Active Users After Twitter Exodus Surge"; Nicholas, "Elon Musk Drove More Than a Million People to Mastodon—But Many Aren't Sticking Around."
28. Rodriguez, "The Bishop and His Star: Citizens' Communication in Southern Chile," 191.

## Chapter 3

1. Technically, there was one more step: the University of Edinburgh's representative to the W3C, Henry Thompson, had to approve Guy's application to join the Social Web Working Group. According to Guy, he was delighted to do so: "Henry Thompson, a professor a floor above me in the building, . . . was like, 'Omigod, a student is interested in W3C! Join whatever groups you want! I'm so happy that anyone cares.' Because, every year, he'd been trying to push the budget to get the university to renew W3C membership, but with no clear reason to justify it to management." Having students, of all people, wanting to join W3C working groups is a good justification for the membership dues.
2. Lemmer-Webber et al., "ActivityPub."
3. The W3C holds copyright but licenses their protocols permissively, granting "Permission to copy, modify, and distribute this work, with or without modification, for any purpose and without fee or royalty," on the condition that those who do so license their work the same way. See "Software and Document License—2015 Version."
4. I should note that some large corporations were involved in the SocialWG, particularly IBM, SAP, and Boeing. These companies had visions of building intracompany social media sites, similar to what Microsoft tried to do with Yammer. This was the dwindling heritage of the OpenSocial Foundation—by the end of that project, it was focusing on intracompany social media. However, multiple interviewees told me that the representatives from IBM and Boeing backed off the SocialWG before it concluded.
5. Russell et al., "The Business of Internetworking," 132.
6. Ibid., 132.
7. To be clear, TikTok wasn't invited, likely because it began in 2016, two years after the start of the SocialWG, and was likely unknown to the SocialWG members at the time.
8. Klemens, "Mastodon—And the Pros and Cons of Moving Beyond Big Tech Gatekeepers."
9. Pierce, "Can ActivityPub Save the Internet?"
10. Gehl, "Alternative Social Media: From Critique to Code."
11. Sevignani, "Facebook vs. Diaspora: A Critical Study."
12. Arrington, "Details Revealed: Google OpenSocial to Launch Thursday."
13. D'Angelo, "What Specific Actions in the Early Years (2004–2007) Led to the Massive Rift Between Facebook and Google?"
14. Grossman, "Mark Zuckerberg—Person of the Year 2010."
15. "OpenSocial Foundation Moving Standards Work to W3C Social Web Activity."
16. I don't want to downplay the achievements of the people involved in the SocialWG, because many of them made excellent, albeit small, alternative social media systems. Prodromou was already well-known for his work on StatusNet and GNU social. Wilkie, another SocialWG member, made a system called Rstatus, which I wrote about in an academic article (Gehl, "Building a Better Twitter"). Christine Lemmer-Webber and Jessica Tallon made Media Goblin. You may not have heard about these before, but as a scholar of alternative social media, I consider the participants in the SocialWG to have excellent resumes. The issue was that, compared

to corporate social media, these are small and relatively untested systems, making them difficult to draw on for standardization purposes.

17. "Social Web Working Group—Publications."
18. Berners-Lee, "W3C NOTE Process." I should point out that producing an ActivityPub note instead of a standard was entirely possible—in addition to producing seven standards, the SocialWG also produced four notes.
19. The exact language in the SocialWG Charter is this: "the Working Group may not develop the Federation Protocol into a Recommendation and it may become a Working Group Note." Translation: the group might write some ideas down about federation, but if there is no time, those ideas would be published as a mere note, on par with a mere commentary—decidedly not a W3C standard. As Bengo, a SocialWG member told me, this may have been due to the idea that the federation protocol was seen as relying on the social syntax and the social API, so they had to make those first. See "Social Web Working Group Charter."
20. "Socialwg/2015-12-02-Minutes—W3C Wiki."
21. The original image posted by Kevin Marks is no longer online, but it is cached here: https://us-browse.startpage.com/av/anon-image?piurl=http%3A%2F%2Fi.imgur.com%2F9nOuB7F.jpg&sp=1697213717T257013b42140cb5a8d8a08418aa3cab8dadaec99225be906f97e170883d462e4. For a discussion of the "Stop Trying to Make X Happen" meme, see "Stop Trying to Make Fetch Happen."
22. "Socialwg/2015-12-02-Minutes—W3C Wiki."
23. Tiara, "Mastodon 101."
24. Ratta, "Digital Socialism Beyond the Digital Social," 104.
25. ghost, "Custom Federation Levels (at the Very Least, for Private Posts) · Issue #712 · Mastodon/Mastodon"; Trev, "GNU Social, Pleroma, and the Mastodon Culture Conflict."
26. Trev, "GNU Social, Pleroma, and the Mastodon Culture Conflict."
27. For a discussion of visibility and violence, see Ratta, "Digital Socialism Beyond the Digital Social," 104.
28. I have not found a specific moment when Mastodon developers learned about ActivityPub, but there were many points of contact between the projects. Evan Prodromou mentioned that Mastodon developers were at least lurking in the SocialWG meetings. SocialWG member Wilkie was active in Mastodon's Github in 2017, providing a link between the projects. Christine Lemmer-Webber worked with Rochko and Mastodon developer Nightpool in that period, as well. Rochko recalls working with Lemmer-Webber early on. In addition, Lemmer-Webber recalls joining Mastodon during its early days and talking about how ActivityPub could help it. See Lemmer-Webber et al., "Open Social Media: Origin Stories"; Tilley, "One Mammoth of a Job."
29. There are least two key moments in this conflict. In mid-2016, Lemmer-Webber advocated for OStatus-like capability that would enable addressing messages to specific groups, a capability that would eventually be included in ActivityPub. Çelik argued against this, saying it was too hard: "If privacy were easy, then OStatus would have done it." In another meeting later that year, Lemmer-Webber would again call for privacy controls, while Çelik would argue it was out of scope: "This is a new topic." The conflict between Lemmer-Webber and Çelik is palpable, visible even in the relatively dull meeting minutes of the SocialWG. See "Socialwg/2016-06-07-Minutes—W3C Wiki"; "Socialwg/2016-11-17-Minutes—W3C Wiki."
30. Lemmer-Webber and Lemmer-Webber, "Fediverse Reflections While the Bird Burns."
31. Mastodon's developers, including Eugen Rochko, were discussing adopting ActivityPub in April 2017. The actual implementation happened later that year in September. See strugee, "ActivityPub Support · Issue #1557 · Tootsuite/Mastodon"; Rochko, "Mastodon and the W3C."
32. The address for the W3C's Mastodon server is now w3c.social.
33. I reached out to Çelik for an interview but did not hear back.
34. "Socialwg/2017-12-05-Minutes—W3C Wiki."
35. Kazemi, "A Highly Opinionated Guide to Learning About ActivityPub."
36. Lemmer-Webber et al., "ActivityPub."
37. Ibid.
38. "Differences in Mastodon API Responses from Vanilla Mastodon—Pleroma Documentation"; "REST API Quick Start | PeerTube Documentation"; "Apps API Methods—Mastodon Documentation"; "API."
39. Kazemi, "A Highly Opinionated Guide to Learning About ActivityPub."
40. tedu, "ActivityPub as It Has Been Understood."

41. Lemmer-Webber, "Split Off ActivityPub CG or WG," October 1, 2023.
42. Guy, "The Presentation of Self on a Decentralised Web," chap. 1.
43. Ibid.
44. Ibid.
45. Ibid.

## Chapter 4

1. Research on furry culture is scarce. What little of it exists tends to focus on it purely as a sort of sexual deviancy. The most conscientious research on the culture is being conducted by Sharon Roberts of the University of Waterloo, as well as a team of people called the International Anthropomorphic Research Project. See https://furscience.com for more information.
2. "Neuromatch Social." They formalized their code of conduct in the months since this post.
3. Complicating this a bit: many Mastodon and fediverse instances do, in fact, use the term "Terms of Service," but even in those cases, it's not quite the same as what we're used to in corporate social media. These Mastodon instances "Terms" read and function much more like codes of conduct as described here.
4. This is not to say other, nonfederated online services don't position their users as part of a community. As Josh Braun (2015) documents, the mid-2000s news-sharing social media site Newsvine saw its users as members of a community, complete with a simple, straightforward "Code of Honor" that conditioned moderation. Joseph Reagle's (2010) analysis of Wikipedia's "Five Pillars" also echoes codes of conduct as I describe them here. And Nancy Baym's (2000) analysis of soap opera fandom on Usenet describes that group in terms of "communities of practice." However, as I will show, one factor that sets a Mastodon or fediverse code of conduct apart is its role in aiding instance-to-instance relations. In that case, the history of another previous system, Bulletin Board Services, may be instructive. See Braun, *This Program Is Brought to You By—*; Reagle, *Good Faith Collaboration*; Baym, *Tune In, Log On*; Driscoll, *The Modem World.*
5. For example, on the instance I help run, Aoir.social, I joined with the moderators to adopt the Scholar.social code of conduct for our purposes. We went through several revisions. Other instances have engaged in more formal processes to create codes of conduct, such as cooperative governance approaches—see Social.coop and Sunbeam.city for examples.
6. "X Terms of Service," accessed June 30, 2024.
7. There are exceptions, of course. Perhaps the most famous is that of Donald Trump, who was removed from Facebook and Twitter in the wake of his attempted coup of the US government in 2021. However, a few months later, once it was clear Trump would not face immediate legal repercussions and would be free to run for president again, Facebook and Twitter invited him back to their platforms. If you're a bit cynical, you might think that Facebook and Twitter removed him after he was done spending advertising dollars on their platforms, and then invited him back once it was clear he would have millions to spend on ads for a subsequent presidential run—not to mention currying favor if he were to win reelection. Of course, their stated reason is that the presence of an insurrectionist in the public sphere is healthy for a democracy. Let's hear him out, they suggest. Considering what happened in early 2025, the cynical analysis is probably the right one.
8. Kissane and Kazemi, "Findings Report."
9. Finley, "The Woman Bringing Civility to Open Source Projects."
10. Coleman, *Coding Freedom*, 48.
11. See, for example, the archived timeline here: https://geekfeminism.fandom.com/wiki/Timeline_of_incidents.
12. Perhaps the most exhaustive documentation of the lack of diversity in the free and open-source sector is Dunbar-Hester, *Hacking Diversity*.
13. ashe dryden [@ashedryden], "I Will Not Attend or Speak at Conferences or Other Events That Do Not Have a Code of Conduct. #CoCPledge"; Ehmke, "Codes of Conduct."
14. Hightower, "An Environment of Respect."
15. Dunbar-Hester, *Hacking Diversity*.
16. Finley, "The Woman Bringing Civility to Open Source Projects."
17. Coleman, *Coding Freedom*, 106.
18. Raymond, *The Cathedral and the Bazaar*, 89.
19. Ehmke, "Codes of Conduct."
20. For an example of the uneven distribution of resources in the technology sector, see Robert Meija's (2015) analysis of high-speed Internet access in Kansas City, Kansas. Meija studied the

city's program to build venture capital–friendly tech startups, finding that both the infrastructure of connectivity as well as social supports needed to participate in such startups heavily favored white men, even though Kansas City, Kansas, had, at the time of his study, a significant Black population. Mejia, "A Pressure Chamber of Innovation."

21. Cohen, "After Years of Abusive E-Mails, the Creator of Linux Steps Aside."
22. To quickly see pushyocracy in action—and to quickly see how arguments about "the best code" become meaningless—consider so-called religious disputes coders have over even the choice of a coding language. Should this be written in Python? In Perl? In C#? The selection of the coding language doesn't settle matters—there are multiple ways to implement solutions in a chosen language. But for advocates of a particular software language, the stakes appear to be incredibly high, and the Internet is full of flame wars over which language is best. Thus, coding is more about subjective arguments than about objectively great code bubbling to the top, and "communities insufficiently get down to the real business of technical work, and instead rely on flaming to make decisions about which piece of code is worthy of becoming part of the larger software project." Nafus, "'Patches Don't Have Gender,'" 679.
23. Dunbar-Hester, *Hacking Diversity*; Reagle, "Naive Meritocracy and the Meanings of Myth."
24. Robertson, "With Linux's Founder Stepping Back, Will the Community Change Its Culture?"
25. "Contributor Covenant: A Code of Conduct for Open Source and Other Digital Commons Communities."
26. Gargron, "Contributor Covenant Code of Conduct." In my interviews, I asked many people how this came about. None of my interviewees recall it being a debatable point; codes of conduct were something FOSS projects *ought* to include. I take this as evidence of the success of the activism of Ehmke and others in normalizing such a practice by the time Mastodon was being developed.
27. "Mastodon | Awoo.Space | About Text"; fpiesche, "[Feature] Customisable Code of Conduct/Terms of Service Page · Issue #411 · Mastodon/Mastodon"; Rochko, "Write a Terms of Service and Privacy Policy · Issue #115 · Mastodon/Mastodon"; nevillepark, "Code of Conduct · Issue #175 · Mastodon/Mastodon."
28. See https://joinmastodon.org/covenant.
29. One exception to this practice applies to "instances of one," which are Mastodon or fediverse instances with only one user, who also happens to be the admin. Creating a code of conduct for one's self is not seen as necessary. Instead, in those cases, other instances simply observe how that one person acts when making decisions about keeping a connection open.
30. See https://www.loomio.com/socialcoop and https://wiki.social.coop/wiki/Governance.
31. Gehl, "Building a Better Twitter."
32. Even though GNU social somewhat belatedly adopted a Contributor Covenant-style code of conduct, their version also includes the older FOSS libertarian "Code of Conflict," noting that "GNU social has a high submission standard and we want to keep quality code in the codebase and bad code out of it. As such your code will be closely scrutinized, and you might take this criticism personally." Thus, they maintain the older "pushyocracy" model that critics have pointed out. See diogo, "Diogo/Gnu-Social."
33. Those Pleroma instances with codes of conduct are predominantly a fork of Pleroma called Akkoma, which was created by someone who liked Pleroma's features but did not agree with the project's emphasis on free speech absolutism. Complicating matters, *another* fork of Pleroma, called Rebased, seems to be dedicated to free speech absolutism. See hannah, "Akkoma"; Gleason, "Soapbox BE Is Now Rebased."
34. The Free Software Foundation describes four freedoms associated with free software. See https://www.gnu.org/philosophy/free-sw.en.html.
35. This point was stressed to me by Matt Lee during my research for Gehl, "Building a Better Twitter."
36. "Federation, n."
37. Gehl and Zulli, "The Digital Covenant."
38. E.g., see Ladner, "Treaty Federalism"; Young, "Hybrid Democracy: Iroquois Federalism and the Postcolonial Project"; Ladner, "Colonialism Isn't the Only Answer"; Baker, "The Covenantal Basis for the Development of Swiss Political Federalism"; Elazar, "Covenant."
39. See https://www.socialmediaalternatives.org/archive/search?query=mastodon &query_type=keyword &record_types%5B%5D=Item &record_types%5B%5D=File &record_types%5B%5D=Collection &submit_search=Search.
40. See Emsley, "Exploring Mastodon: A Proof of Concept Tool."

41. See https://join-lemmy.org/docs/code_of_conduct.html.
42. See https://github.com/bookwyrm-social/bookwyrm/blob/main/LICENSE.md. Note that "Ethical Source" is the latest project spearhead by Coraline Ada Ehmke and her allies. The Organization for Ethical Source (https://ethicalsource.dev/) is dedicated to developing a legal licensing structure so developers can ensure their software is used in ethical ways, such as not contributing to incarceration or militarization.
43. The current covenantal fediverse replicates many of the community moderation practices Internet historian Kevin Driscoll (2022) finds in his studies of 1980s to 1990s-era bulletin board systems (BBSes). BBSes, like fediverse instances, tend to be run by single figures called "sysops." Sysops had a great deal of power, most especially over the technical details of their boards, but "if they acted like jerks, their users were not likely to return" (143). To maintain their communities, BBS sysops, he argues, "pioneered the difficult practice of community moderation" (143). This includes deciding who should be allowed to join, setting time limits, interpersonal communication, moderating discussions, and in the case of FIDOnet, deciding which other BBSes to pull content from. See *The Modem World*.

## Chapter 5

1. The memoirs appear in a series of blog posts, starting with https://roiskinda.cool/2022/11/recuerda-pv-chapter-1-the-start-part-1.html.
2. Ro, "The Big Fat Intro."
3. Ro, "Recuerda PV," November 24, 2022.
4. Ro, "Recuerda PV," December 2, 2022.
5. Ro, "Recuerda PV," December 9, 2022.
6. Ro, "Dev Diaries 02 & The Numbers for November."
7. Ro, "Dev Diary—Mini Update 02—BIG CHANGES."
8. AdaRoseCannon, "Support for Subscribing to Communal Block Lists · Issue #116 · Mastodon/Mastodon." Also see https://github.com/mastodon/mastodon/issues/116; argxentakato. "Domain Block List Export / Import · Issue #15095 · Mastodon/Mastodon." GitHub, November 4, 2020; https://github.com/mastodon/mastodon/issues/15095; triplefox. "Federated Whitelist · Issue #344 · Mastodon/Mastodon." GitHub, December 8, 2016. https://github.com/mastodon/mastodon/issues/344.
9. Rochko, "I Don't Think Shared Server Blocklists Are a Good Idea." I also interviewed Rochko in November 2023, and while he was more accepting of importable instance blocklists at that point, he repeated similar arguments about their potential downsides.
10. milhouse, "Don't Use Randi Harper's Blocklist. Use Naziblocker Instead"; "The Problem with Personal Block Lists: A Blog."
11. Patel, "Can Mastodon Seize the Moment from Twitter?"
12. Marcia has published a short history of #fediblock. See X, Marcia. "#Fediblock, a Tiny History."
13. "FediBlock."
14. Similar misogynist and racist harassment of a hashtag cocreator has been repeated. Christina Dunbar-Hester's autoethnography of the #asstodon hashtag in 2022 echoes much of what Marcia X experienced in 2018–2020. See Dunbar-Hester, "Showing Your Ass on Mastodon."
15. Jasser et al., "'Welcome to #GabFam,'" 4.
16. It's hard to say with certainty how many people are using Gab. The numbers over the years have varied wildly, prompted in part, it seems, by Torba himself, who has tweeted that millions are using Gab. Some reporting has echoed this claim. However, since Gab is supported by ads, Torba has an incentive to inflate numbers to make it more attractive to advertisers. A 2023 Pew study only considered traffic to Gab, which they put at over 500,000 visitors. Wikipedia, which usually cites sources, claims Gab has 100,000 active users. A recent academic extensive study of Gab by Deghan and Nagappa does not cite or provide any user statistics. Overall, then, I cannot cite any number with any certainty. See Goodwin, "Gab"; "Gab (Social Network)"; Stocking et al., "The Role of Alternative Social Media in the News and Information Environment"; Dehghan and Nagappa, "Politicization and Radicalization of Discourses in the Alt-Tech Ecosystem."
17. Du Mez, *Jesus and John Wayne*, 6.
18. Ibid., 5.
19. Torba, "Responding to Rachel Maddow's Attack on Gab."
20. Torba, "The Path Forward."
21. Torba, "Your Duty to Wake Up Those Around You."

22. Aubin and Stocking, "Key Facts About Gab."
23. Makuch, "The Nazi-Free Alternative to Twitter Is Now Home to the Biggest Far Right Social Network."
24. Torba, "There Is No Freedom of Speech Without Freedom of Reach."
25. Marcuse, "Repressive Tolerance," 85.
26. Those who call for unfettered free speech recognize how repressive tolerance works, which is why they work so hard to flip the narrative upside down. Instead of recognizing that they are in positions of power in places like the United States, white Christian nationalists foment conspiracy theories, such as the idea that trans people (who are a tiny minority in the United States) are everywhere, grooming children and spreading a "trans agenda." In addition, white Christian nationalists also proclaim that they are the victimized minority in the dominant culture, arguing that their ideas are constantly being "canceled." In the case of Torba of Gab, as we have seen, this canceling is done by "Big Tech," which Torba argues is a puppet of Jews. The audacious claim here is that, somehow, tiny minorities (Jews, trans folks) are somehow behind everything and that those in power (predominantly white, Protestant men) are their hapless victims.
27. Caelin, "Decentralized Networks vs. the Trolls," 149.
28. *Gab Employee Explains Why ActivityPub Sucks*; "Rob Colbert on Gab"; "I Believe They Broke . . ."
29. Reuters, "Trump's New App Tops Downloads in Apple Store"; Goldstein and Mac, "Trump's Truth Social Is Poised to Join a Crowded Field."
30. Cherry, *The Case for Rage*, 43–44.
31. Ibid., 44.
32. Flowers, "The Whiteness of Mastodon."
33. This echoes what Alice Marwick finds in her study of harassment, where intersectional identities provide multiple "attack vectors" for harassers. See Marwick, "Morally Motivated Networked Harassment as Normative Reinforcement."
34. Flowers, "The Whiteness of Mastodon."
35. Hanhardt, *Safe Space.*
36. Marcia expands on these points in a recent interview. "What took me aback regarding the fediverse is that my networks were mostly 'leftists' and self-proclaimed radical thinkers regarding race, ableism, gender, patriarchy, sexuality, et cetera, and yet what I was being exposed to was a lot of naiveté or hostility for questioning whiteness as a basis for many people's takes or approaches to these subject matters. And if I were to question or push back on their whiteness, I was often accused of being biased myself." See X and Kiam, "Blackness in the Fediverse: An Interview with Artist Marcia X."
37. https://tweaking.thebad.space/about.
38. Pulliam, "Federation Safety Enhancement Project (FSEP)."
39. See https://github.com/iftas-org/resources/blob/main/DNI/dni.csv for the IFTAS list and IFTAS.org for the organization's website.
40. Oliphant, "Oliphant.social Mastodon Blocklists."
41. This is not to say individuals on AoIR.social have not been harassed, just that I have not received such a report as an admin—I have not been asked to block instances by AoIR.social members. We do get the occasional spam, but these tend to be innocuous, not racist or transphobic.
42. Bowker and Star (1999) make a similar point in their discussion of the Yellow Pages in phone books. "Under the community events section in the beginning, next to the Garlic Festival and the celebration of the anniversary of the city's founding, the Gay and Lesbian Pride Parade is listed as an annual event. Behind this simple telephone book listing lie decades of activism and conflict—for gays and lesbians, becoming part of the civic infrastructure in this way betokens a kind of public acceptance almost unthinkable thirty years ago." See *Sorting Things Out: Classification and Its Consequences*, 53.
43. Cherry, *The Case for Rage*, 5.
44. The image, which appeared on an earlier version of The Bad Space, is from Ussama Azan, found on Unsplash: https://unsplash.com/photos/935EcPU1pBI.
45. Brock, *Distributed Blackness.*
46. "In spite of it all" is a term Johnathan Flowers uses to discuss the rise of Black Twitter. In spite of private ownership and its imbrication in structures of oppression, Flowers notes Black Twitter took the affordances of Twitter to create something new and joyful. Likewise, I would suggest that Ro sees the fedi's possibility—in spite of it all. See Flowers, "The Whiteness of Mastodon."
47. Brock, *Distributed Blackness*, 6.

## Chapter 6

1. This is often attributed to Andrew Lewis or Douglas Rushkoff in the early 2010s. However, the underlying idea predates the Internet. Perhaps the most in-depth analysis of how audiences are the product, rather than the consumer, of advertising-funded media comes from Canadian media theorist Dallas Smythe, who argued that the audience does the work of watching advertisements, helping capitalist firms realize profits by purchasing the items that are advertised. See Smythe, "On the Audience Commodity and Its Work."
2. Fuchs, "The Political Economy of Privacy on Facebook"; Deibert, "The Road to Digital Unfreedom"; Mejias and Couldry, *The Costs of Connection*.
3. Monge Roffarello and De Russis, "Towards Understanding the Dark Patterns That Steal Our Attention."
4. Fuchs, *Critical Theory of Communication*, 134.
5. Zuboff, *The Age of Surveillance Capitalism*.
6. Jefferson Cord, as quoted in Gehl, "'Why I Left Facebook': Stubbornly Refusing to Not Exist Even After Opting Out of Mark Zuckerberg's Social Graph," 228.
7. "Google Buys YouTube for $1.65 Billion."
8. Conger and Hirsch, "Elon Musk Completes $44 Billion Deal to Own Twitter."
9. "Meta Reports Fourth Quarter and Full Year 2022 Results."
10. "Alphabet Announces Fourth Quarter and Fiscal Year 2022 Results."
11. Many people who think this way forget the existence of Wikipedia, which is a nonprofit and yet is one of the most visited websites in the world. As I wrote in my first book, Wikipedia also started as a classic surveillance capitalist project. Their intention was to sell ads around the user-generated articles. However, a labor strike by Wikipedia editors in Spain compelled the founders to switch to a nonprofit, donation-funded model. See chapter 5 of Gehl, *Reverse Engineering Social Media: Software, Culture, and Political Economy in New Media Capitalism*; also see Enyedy and Tkacz, "'Good Luck with Your WikiPAIDia': Reflections on the 2002 Fork of the Spanish Wikipedia."
12. Fisher, *Capitalist Realism*.
13. Gibson-Graham, "Diverse Economies," 615.
14. White and Williams, "Anarchist Economic Practices in a 'Capitalist' Society: Some Implications for Organisation and the Future of Work," 961.
15. Gibson-Graham, "Diverse Economies," 627.
16. White and Williams, "Anarchist Economic Practices in a 'Capitalist' Society: Some Implications for Organisation and the Future of Work," 953.
17. Full disclosure: for several years, Scholar.social was my home on the fediverse, and I am currently donating a small amount of money each month. I disclosed this to Socrates when I interviewed him.
18. corbet, "The Open Collective Foundation Is Shutting down [LWN.Net]." It is possible some other organization will pick up the slack—several people have informally talked to me about this—but as of this writing, things are uncertain.
19. "Chaos.Social Info."
20. The name is spelled "Hueristic" purposefully. As of this writing, Ro's Hueristic Instruments is in the early stages. Most of the information about the project can be found at the Mastodon instance: https://h-i.social/about.
21. In April 2024, at the same time as he announced Mastodon's new US nonprofit, Rochko announced that Mastodon had lost its gGmBH status in Germany for reasons that were not communicated to him. However, as of this writing (July 2024), the Mastodon organization website refers to it as "Mastodon gGmBH." I have not seen reporting that Mastodon has regained gGmBH status in Germany. See Rochko, "Mastodon Forms New U.S. Non-Profit."
22. "Mastodon Annual Report 2022."
23. White and Williams, "Anarchist Economic Practices in a 'Capitalist' Society: Some Implications for Organisation and the Future of Work," 957.
24. Spade, "Solidarity Not Charity," 136.
25. Full disclosure: I'm one of the folks who regularly boost these posts.
26. Spade, "Solidarity Not Charity"; Brice, "Geographies of Vulnerability"; Mould et al., "Solidarity, Not Charity."
27. Spade, "Solidarity Not Charity," 137.
28. Mould et al., "Solidarity, Not Charity," 874.
29. Baldwin, *The Fire Next Time*.

30. Spade, "Solidarity Not Charity," 137.
31. Mould et al., "Solidarity, Not Charity," 875.
32. Petre et al., "'Gaming the System.'"
33. Duffy et al., "Platform Practices in the Cultural Industries."
34. *Pepper & Carrot.*
35. Revoy, "Pepper & Carrot (MOTION COMIC) Episode 6."
36. Revoy, "License F.A.Q."
37. Revoy, "The Derivative Pepper&Carrot Motion Comic Project by Nikolai Mamashev—David Revoy."
38. ActivityPub is licensed under a permissive W3C license. See https://www.w3.org/copyright/software-license-2015/. Mastodon is licensed with the AGPL 3.0 license. See https://github.com/mastodon/mastodon/blob/main/LICENSE. Both licenses allow others to use and modify the works, though they do include some restrictions. For example, in the case of the AGPL, requirements include any modified version must make the source code available to the public. This requirement led to a threat of legal action against Donald Trump's Truth Social, which took Mastodon's code but did not make Truth's modified source code publicly available. See Kuhn, "Trump's Social Media Platform and the Affero General Public License (of Mastodon)."
39. Sinnreich, *Mashed Up*, chap. 5.
40. The need for algorithmic valorization leads to much effort by artists to "game" algorithmic systems. In response, corporate platforms arbitrarily condone or halt such efforts to appear to be neutral channels. See Petre et al., "'Gaming the System.'"
41. Duffy, "The Romance of Work."
42. Terranova, "Free Labor: Producing Culture for the Digital Economy."
43. Striphas, "Algorithmic Culture"; Hallinan and Striphas, "Recommended for You."
44. Striphas, "Algorithmic Culture."
45. For an analysis of why we need alternatives for artists, see Poell, "Three Challenges for Media Studies in the Age of Platforms."
46. For example, as of this writing in mid-2024, a major concern among creators on the fediverse is a recent change in Patreon prompted by Apple, where Apple is demanding a 30% cut of subscription revenues in exchange for Patreon maintaining an Apple app. See Heydari, "Apple Looks for More Money by Changing the Math on Patreon."
47. For academic works that analyze content moderation, see Roberts, *Behind the Screen*; Gibson, "Free Speech and Safe Spaces"; Gillespie, *Custodians of the Internet*; Matias, "The Civic Labor of Volunteer Moderators Online."
48. Spoons are metaphorical units of energy. The metaphor comes from a blog post about living with a chronic disease. See Miserandino, "But You Don't Look Sick?"
49. Valens, "Mastodon Is Crumbling—And Many Blame Its Creator."
50. Chan, "Open-Source Developers Are Burning Out, Quitting, and Even Sabotaging Their Own Projects—And It's Putting the Entire Internet at Risk," March 19, 2022.
51. The practice of having open registrations on the Mastodon.social and Mastodon.online instances is an example of the tremendous pressure faced by Rochko and the Mastodon gGmBH project. Multiple admins I spoke to think open registrations is a mistake, because it allows spammers and trolls to freely sign up. They argue that Mastodon.social, in particular, is too large and not well moderated. Indeed, as an instance admin myself, the most common problem I've faced is spam coming from Mastodon.social, and I've often limited that server to cut down on spam. In light of this, Rochko's critics argue that Mastodon.social should close registrations, so that would-be new Mastodon members would be required to spread out across smaller instances. This would not only improve moderation on the fediverse but also prevent any given Mastodon instance from being too dominant.

    However, given that would-be new Mastodon members must go *somewhere*—they must pick an instance to join—that many Mastodon instances limit registrations in order to stay small, and that people who are used to centralized systems do not understand how instances work, Rochko argues that he must keep Mastodon.social open as a default option.

    As he told me in an interview, "[Mastodon.social is from] a trustworthy company, probably going to be around for a long time. You don't necessarily have to trust the individuals involved in running it as long as you trust the company because they hire people, they fire people—you know, whatever, right? It's a company. It's not going to disappear. It's not going to get into fights with other little servers unnecessarily. Moderation's gonna be good. So it's just going to be more or less traditional social media service that you'd expect."

Having a large instance ready for new member sign-ups replicates the ease of signing up for Twitter/X while at the same time bringing people into noncentralized social media. Otherwise, the network would be perceived as confusing and closed off. This is but one of many difficult trade-offs Rochko and the paid Mastodon developers must make.

52. Times, "Twitter Rival Mastodon Rejects Funding to Preserve Nonprofit Status."

## Chapter 7

1. Rixen et al., "The Loop and Reasons to Break It."
2. "Meta Reports Fourth Quarter and Full Year 2022 Results."
3. Brandon, "Instagram Is Outpacing TikTok for User Growth. Here's Why."
4. Mehta, "Google Says YouTube Shorts Has Crossed 50 Billion Daily Views."
5. Binswanger, "Is There a Growth Imperative in Capitalist Economies?"; cf. Gordon and Rosenthal, "Capitalism's Growth Imperative."
6. In computer network diagrams, clouds indicate a portion of the network that you don't need to think about. In the early days of corporate networks, organizations ran all their own computing infrastructure on premises (or "on prem" in network-speak). In such a system, there are no clouds in the network diagram—the local IT folks knew all the network details. "Cloud infrastructure" arose as a service to organizations who wanted to move their computer resources "off prem" to third-party providers, such as Amazon and Microsoft. At that point, an organization's IT folks would put a picture of a cloud on their network diagram, indicating they didn't know what the actual infrastructure was—those resources are Amazon's or Microsoft's problem.
7. Monserrate, "The Staggering Ecological Impacts of Computation and the Cloud."
8. Siddik et al., "The Environmental Footprint of Data Centers in the United States"; also see Hogan, "Data Flows and Water Woes" for a study of government data storage in desert environments. However, measuring exact impacts of server farms remains difficult due to the scale and competing metrics. See Pasek et al., "The World Wide Web of Carbon."
9. Maxwell and Miller, "Environmental Materialism and Media Globalization."
10. See https://www.esa.int/Space_Safety/Space_Debris/Space_debris_by_the_numbers.
11. Pasek, "On Being Anxious About Digital Carbon Emissions," 1.
12. Laser et al., "The Environmental Footprint of Social Media Hosting"; also see Stevenson and Valente Pinto, "Distinction and Alternative Tech," 11.
13. Paulson, "Economic Growth Will Continue to Provoke Climate Change."
14. Hickel et al., "Degrowth Can Work—Here's How Science Can Help," 401.
15. Zuboff, *The Age of Surveillance Capitalism.*
16. Sayedi, "Real-Time Bidding in Online Display Advertising."
17. Stiegler, *For a New Critique of Political Economy.*
18. Facebook whistleblower Frances Haugen testified to US Congress about the intensification of engagement with social media. As she noted, Facebook made changes to its algorithms around 2018 that led to more polarizing, inflammatory content being emphasized, leading to the spread of hate speech. I want to thank Josh Braun, one of the reviewers of this manuscript, for reminding me of this. See Ordonez, "Key Takeaways from Facebook Whistleblower Frances Haugen's Senate Testimony"; "The Facebook Files, Part 4"; Maher, "Facebook's Algorithm Comes Under Scrutiny."
19. Raman et al., "Challenges in the Decentralised Web," 222. This study, however, is problematic. The authors' dataset includes a reported 67 million posts from the fediverse, drawn from the now-defunct Mastodon Monitoring Project. There are ethical issues here: on the fediverse, the culture of consent is very strong, so much so that studies like this are condemned for taking people's posts without their consent.
20. See the Fediverse Observer map at https://fediverse.observer/map.
21. Han et al., "On the Infrastructure Providers That Support Misinformation Websites"; Delfanti, "Machinic Dispossession and Augmented Despotism."
22. Full disclosure: I know Roel quite well. I served as a 30% reader of his dissertation project at Malmö University in Sweden. In the Swedish system, PhD students operate in stages, with external reviewers contributing comments and critiques. I now have no influence over the remaining stages of his PhD project. However, we have collaborated in writing about the fediverse. See Abbing and Gehl, "Shifting Your Research from X to Mastodon?"
23. Roscam Abbing, "'This Is a Solar-Powered Website, Which Means It Sometimes Goes Offline,'" 3.
24. Wajcman, *Pressed for Time,* 20.
25. Ojala et al., "Anxiety, Worry, and Grief in a Time of Environmental and Climate Crisis."

26. Flynn, "Solarpunk: Notes Toward a Manifesto."
27. Reina-Rozo, "Art, Energy and Technology," 50. Arguably, solarpunk's roots run deeper than the 2010s, with Kim Stanley Robinson's 1990s *Mars* trilogy and Marge Piercy's 1976 book *Woman on the Edge of Time* as representatives of the genre. Robinson's *Mars* series features humans terraforming (or aereoforming) Mars. During the process, they develop a new ecological economics capable of resuscitating the damaged Earth they had left behind. Piercy's book contemplates a future where genders are fluid and people draw on Indigenous American knowledge to live closely with the land and the sun.
28. These themes appear across multiple stories in the collection by Wagner and Wieland, *Sunvault*.
29. Mirov, "The Desert, Blooming," 114.
30. Reina-Rozo, "Art, Energy and Technology."
31. Ibid.
32. Watson, "The Boston Hearth Project."
33. Hickel et al., "Degrowth Can Work—Here's How Science Can Help," 402.
34. Yaszek, "Afrofuturism, Science Fiction, and the History of the Future."
35. Reina-Rozo, "Art, Energy and Technology," 50.
36. Laser et al., "The Environmental Footprint of Social Media Hosting."
37. E.g., Pasek et al., "The World Wide Web of Carbon."
38. Pasek, "On Being Anxious About Digital Carbon Emissions," 3.
39. Available at https://solar.lowtechmagazine.com/.
40. Decker, "About the Solar Powered Website."
41. Allebach and Stradling, "Computer-Aided Design of Dither Signals for Binary Display of Images."
42. See https://www.blackholeberlin.com/en for details. The use of AI art is particularly galling here, since research is showing that contemporary machine learning–based media generation is environmentally damaging. See Li et al., "Making AI Less 'Thirsty.'"
43. Hickel et al., "Degrowth Can Work—Here's How Science Can Help," 403.

## Chapter 8

1. Choudhury and Sn, "Exclusive"; Ray, "Meta Is Creating a Decentralized Twitter Alternative Reportedly Codenamed 'P92'"; Newton, "Meta Is Building a Decentralized, Text-Based Social Network."
2. Threads was initially developed under the code names "Project 92" and "Barcelona."
3. Heath, "This Is What Instagram's Upcoming Twitter Competitor Looks Like."
4. pierat, "Meta/Facebook Is Inviting Fediverse Admins Under NDA for 'Meetings.'"
5. Available online at fedipact.online.
6. "#Fedipact—The Instances Blocking Zuckerberg's Threads.Net."
7. "♥ Why? ♥."
8. Free Fediverse, "a resource for those working to save the fediverse network of online communities from absorption into surveillance capitalism," is collecting citations about "Meta's crimes against humanity" at this page: https://freefediverse.org/index.php/Nightmares. The page includes stories about the genocide in Myanmar, misinformation and conspiracy theories, and purposely addictive design, among many others.
9. "♥ Why? ♥". The website links to the following news stories: Mozur, "A Genocide Incited on Facebook, with Posts from Myanmar's Military"; Booth, "Facebook Reveals News Feed Experiment to Control Emotions"; Mac and Frenkel, "Internal Alarm, Public Shrugs." Fediverse members have also written at length about these issues. See Seirdy, "De-Federating P92"; Kissane, "Untangling Threads."
10. Musambi, "Kenyan Facebook Moderators Accuse Meta of Not Negotiating Sincerely"; Siele, "'It's Been Tough for Us.'"
11. Vezina, "Defederate Meta."
12. Frenkel, "QAnon Is Still Spreading on Facebook, Despite a Ban."
13. Nix, "Facebook Bans Hate Speech but Still Makes Money from White Supremacists."
14. Revoy, "Don't Roll Out the Red Carpet for Them."
15. "#Fedipact—The Instances Blocking Zuckerberg's Threads.Net."
16. Quirk, "Threads and the Fediverse."
17. Gruber, "More on Preemptively Blocking Facebook's Imminent ActivityPub Entry."
18. The original post, from John Gruber, appeared on Bluesky, another Twitter alternative. For a screenshot, see "Misfit Loser Zealot Club."

19. Davis, "Threads Is Officially Starting to Test ActivityPub Integration."
20. Jones and Diaz, "5 Things to Know About Meta's Threads App Before You Entangle Your Instagram Account."
21. Porter, "Instagram's Threads Surpasses 100 Million Users." See the Introduction, note 17 for sources for calculating the size of the fediverse.
22. Ernst, "Congratulations!" December 15, 2023. However, as of March 2024, Evan Prodromou (one of the chairs of the Social Web Incubator Community Group, the group charged with maintaining ActivityPub) has reported to me that Meta is not involved in ActivityPub development. Prodromou believes Meta representatives may not feel welcome in the SWICG due to anti-Meta sentiments expressed on the mailing list. "I would like to see them get involved," he told me, especially because Meta could provide resources to the group, including funding for group members to meet in person.
23. Rochko, "My Wife Is Looking Forward to Deleting Her Instagram Account Once She Can Connect with the Same Folks from Her Mastodon Account. Being Able . . ."
24. Chris Messina, quoted in Gruber, "More on Preemptively Blocking Facebook's Imminent ActivityPub Entry."
25. Quirk, "Threads and the Fediverse."
26. Rochko, "What to Know About Threads."
27. Trottier, "Federation with #Meta Actually Hurts Meta."
28. Knight, "Meta Doesn't Need ActivityPub to Slurp Up Your Data."
29. Claburn, "Meta, Which Pays for Web Scraping, Sues to Stop Web Scraping"; Feathers et al., "Facebook Is Receiving Sensitive Medical Information from Hospital Websites—The Markup"; Morrison, "Meta Is Getting Data About You from Some Surprising Places"; Wagner, "This Is How Facebook Collects Data on You Even If You Don't Have an Account"; Cuthbertson, "Project Ghostbusters."
30. Chambers, "Project92 and the Fediverse—A Smarter Battle Plan to Protect the Open Social Web."
31. Nafus, "'Patches Don't Have Gender.'"
32. Marcuse, "Repressive Tolerance."
33. The Mastodon server covenant's first commitment is to "Active moderation against racism, sexism, homophobia and transphobia. Users must have the confidence that they are joining a safe space, free from white supremacy, anti-semitism and transphobia of other platforms." The argument vanta and others make is that Meta has repeatedly fallen short of that commitment. See https://joinmastodon.org/covenant.
34. Fediverse developer Darius Kazemi's "Run Your Own Social" guide jokes, "'It's hard work but it's worth it' [is] an ongoing theme" in federated social media. See Kazemi, "How to Run a Small Social Network Site for Your Friends."
35. Elazar, "Covenant," 2; also see Elazar, "Confederation and Federal Liberty."
36. Elazar, "Confederation and Federal Liberty," 4.
37. Similarly, Indigenous scholar Kiera Ladner has studied Indigenous governance structures, which demand what she calls "treaty federalism," a sovereign, nation-to-nation relationship between Indigenous nations and colonial governments, such as that of Canada. Structurally, the relationship is shaped by covenants between groups. See Ladner, "Treaty Federalism."
38. This echoes what Iris Marion Young argues for when she argues for "decentered federalism." See Young, "Hybrid Democracy: Iroquois Federalism and the Postcolonial Project."
39. Prodromou, "Big Fedi, Small Fedi." Prodromou puts this individual-centric perspective under a broader umbrella he calls "Big Fedi," which is an approach to federation that seeks growth and thus welcomes commercial services. He contrasts this with "Small Fedi," where instances are central units and commercial services are shunned.
40. Porter, "Threads Launches for Nearly Half a Billion More Users in Europe."
41. Many digital media companies have used a similar tactic when faced with potential or actual state regulation. For example, the online advertising industry has faced potential privacy regulations in the United States. Their response is to create systems where end users could "opt out" of behavioral tracking. However, these systems are largely for show and are hardly functional. See Gehl, "The Politics of Punctualization and Depunctualization in the Digital Advertising Alliance."
42. Wagner, "Musk's X 2023 Ad Sales Projected to Slump to About $2.5 Billion"; Mac and Conger, "X May Lose Up to $75 Million in Revenue as More Advertisers Pull Out"; Barwick, "X Ad Costs and Engagement Continue to Fall, per Report."

43. This point is also made by Vezina, "Defederate Meta."
44. E.g., Vezina, although multiple people have also mentioned this in various fediverse posts. The way this would work might be like this: if I am on a fediverse instance that federates with Threads, and I post something that Threads's algorithm boosts to Threads users, who in turn boost, like, or comment upon the post, then my post will appear as more important (in terms of boosts, likes, or comments) to the rest of the fediverse than non-algorithmically boosted posts. Such a shift may increase what Roth and Lai call the "attack surface" of the fediverse, enabling more "inauthentic engagement and behavioral manipulation." See Roth and Lai, "Securing Federated Platforms," 20.
45. Stray et al., "The Algorithmic Management of Polarization and Violence on Social Media."
46. This is according to FediDB, which has been tracking #fedipact. See https://fedidb.org/current-events/anti-meta-fedi-pact, last accessed March 20, 2024.
47. The Fedi Garden rules are available here: https://fedi.garden/about-this-site/.
48. Prodromou, "Big Fedi, Small Fedi."
49. Perez, "Tumblr's 'Fediverse' Integration Is Still Being Worked on, Says Owner and Automattic CEO Matt Mullenweg."
50. Stubblebine, "Medium Embraces Mastodon."

## Chapter 9

1. Specifically, I focused very heavily on Marxian political economy in my early works, which enabled me to consider how attention may be commodified but often blinded me to problems of content moderation and community management.
2. Gilligan, *In a Different Voice*; Noddings, *Caring, a Feminine Approach to Ethics & Moral Education*; Fisher and Tronto, "Towards a Feminist Theory of Caring."
3. Fisher and Tronto, "Towards a Feminist Theory of Caring," 40.
4. Ibid., 56.
5. Kittay, "The Ethics of Care, Dependence, and Disability."
6. Bass, "When Care Trumps Justice"; Raghuram, "Race and Feminist Care Ethics"; Elliott, "Caring Masculinities"; Kittay, "The Ethics of Care, Dependence, and Disability"; Martin et al., "The Politics of Care in Technoscience"; Mattern, "Maintenance and Care."
7. These four phases of caring come from an influential essay on the ethics of care: Fisher and Tronto, "Towards a Feminist Theory of Caring."
8. One of the origin stories of the ethics of care comes from a debate between a Harvard psychologist, Lawrence Kohlberg, and his PhD student, Carol Gilligan. According to philosopher Maureen Sander-Staudt, "Kohlberg had posited that moral development progressively moves toward more universalized and principled thinking and had also found that girls, when later included in his studies, scored significantly lower than boys. Gilligan faulted Kohlberg's model of moral development for being gender biased, and reported hearing a 'different voice' than the voice of justice presumed in Kohlberg's model." See Sander-Staudt, "Care Ethics."
9. van der Vossen and Christmas, "Libertarianism."
10. Wessalowski and Karagianni, "From Feminist Servers to Feminist Federation," 195.
11. Raghuram, "Race and Feminist Care Ethics," 614.
12. Mol et al., "Care: Putting Practice into Theory," 13.
13. Mattern, "Maintenance and Care."
14. Martin et al., "The Politics of Care in Technoscience," 627.
15. X and Kiam, "Blackness in the Fediverse: An Interview with Artist Marcia X," 47.
16. The collectively run Mastodon instance Social.coop explicitly recommends rotating content moderators to avoid burnout. "Make sure that people doing moderation work have some training and support," they suggest. "Ensure that the role rotates, so people are less likely to burn out." See "How to Make the Fediverse Your Own."
17. The Mastodon Server Covenant recognizes this practice, requiring adherents to provide at least three months' notice before shutting down. "Sometimes services shut down, it is the cycle of life," the Covenant states. "But users must have the confidence that their account will not disappear overnight, so that they have time to export their data and find another server." See https://joinmastodon.org/covenant.
18. Marcuse, "Repressive Tolerance"; Fopp, "'Repressive Tolerance.'"
19. Perhaps the most influential care ethicist who discusses reciprocity is Nel Noddings. See Noddings, *Caring, a Feminine Approach to Ethics & Moral Education.*

20. "The Problem with Personal Block Lists: A Blog."
21. Raghuram, "Race and Feminist Care Ethics," 623.
22. I take this last point from Allie Hart's "Mourning Mastodon" article, which appeared early in Mastodon's history and documents how queer, Black, and disabled folks felt ignored by Rochko, unless they "constantly engaged, consciously or not, in a battle for recognition and validation, both of themselves (who were frequently left out of the narrative when it came to Mastodon development) and of their needs and ideas." See Hart, "Mourning Mastodon."
23. Gehl and Zulli, "The Digital Covenant," 13.
24. Raghuram, "Race and Feminist Care Ethics," 624.
25. Roberts, *Behind the Screen*; Gillespie, *Custodians of the Internet*; Dunbar-Hester, *Hacking Diversity*; Kalbag, *Accessibility for Everyone*.
26. Eva Feder Kittay, as quoted in Sander-Staudt, "Care Ethics."
27. See https://fediverse.observer/map; last visited February 29, 2024.
28. Robertson, "OLPC's $100 Laptop Was Going to Change the World—Then It All Went Wrong."
29. Science and technology studies scholar Madeleine Akrich notes that technologists often design solutions with an ideal user and situation in mind, a process she calls "de-scribing." When the technology finally makes its way to the target situation, it fails, often for simple and predictable reasons typically having to do with the lack of infrastructure and basic necessities. See Akrich, "The De-Scription of Technical Objects."
30. Gow, "Alternative Social Media for Outreach and Engagement."
31. Ibid.
32. Card, "Caring and Evil," 102.
33. Mol et al., "Care: Putting Practice into Theory," 13.
34. Ibid., 13–14.
35. Baldwin, *The Fire Next Time*, 88, 92.

## Epilogue

1. This is similar to how media studies scholar Christina Dunbar-Hester has engaged with the fediverse: an attempt to inject some levity. See Dunbar-Hester, "Showing Your Ass on Mastodon."
2. For a list of films that Monsterdon participants have watched, see https://letterboxd.com/justcheri/list/monsterdon/.
3. Gibron, "Depth of Field."
4. Rose, "Mining Memories with Donald Trump in the Anthropocene"; Klein, "Toxic Nostalgia, From Putin to Trump to the Trucker Convoys"; Hassler-Forest, "'When You Get There, You Will Already Be There' Stranger Things, Twin Peaks and the Nostalgia Industry."
5. Chatterjee, "Social Media's Fragmented Future"; Paul, "Is Threads the Good Place?"
6. Paul, "We Tried Threads, Meta's New Twitter Rival. Here's What Happened."
7. Clouse, "Threads."
8. Ford, "On Threads, SPILL, and Social Media Being 'Cool' Again."
9. While it is now distinct from Twitter, its roots are deeply tied to that company. Bluesky was initially conceived of by Dorsey and developed by Twitter under the leadership Parag Agrawal, who would go on to be CEO of Twitter. It was later spun off Twitter as a separate company, headed by Jay Graber. Bluesky is developing their own technical protocol, AT Protocol, which they intend to use to decentralize Bluesky into a network of servers. In addition, Bluesky has stated their intention to turn AT Protocol over to the stewardship of the Internet Engineering Task Force, a standards body. However, at the time of this writing, there is only a single Bluesky instance, bsky.app, and the protocol is still only controlled by the Bluesky corporation. Not much, if any of it, has been decentralized.
10. Karpf, "Bluesky Is Just Twitter, Without the Burden of Everything Ruining Twitter."
11. Greenwood, "How I Accidentally Ruined Bluesky with Pictures of Sexy Alf."
12. Knibbs, "I Regret to Inform You That Bluesky Is Fun."
13. See https://docs.castopod.org/ for an implementation of podcasting on ActivityPub; https://github.com/H4kor/fedi-games for a project using ActivityPub for game play.
14. In early 2024, I attended a Decentralized Social Media Workshop at Princeton University, which was attended by fediverse and Bluesky developers, representatives from various digital media nonprofits, and academics. One of the more pressing concerns for this group was what to do about disinformation in federated social media.
15. Cole, "Switter, the Twitter for Sex Workers, Is Shutting Down."

## Appendix

1. However, the field is growing, especially in the past few years. I have collected a bibliography of alternative social media research at https://socialmediaalternatives.org/bib.html.
2. Gehl, *Weaving the Dark Web: Legitimacy on Freenet, Tor, and I2P*.
3. Wähner et al., "'Don't Research Us'—How Mastodon Instance Rules Connect to Research Ethics."
4. Molyneux and McGregor, "Legitimating a Platform," 1578.
5. In this sense, I could argue that taking posts out of context violates their contextual integrity, to use a concept from Helen Nissenbaum (2004). However, journalists have often conceived of Twitter/X as a public forum, using that as a justification for taking posts out of context. I do not think we can make the same case for the fediverse, given that many of its members have joined it to avoid such violations of contextual integrity. See Nissenbaum, "Privacy as Contextual Integrity."
6. Ess and AoIR Ethics Working Committee, "Ethical Decision-Making and Internet Research: Recommendations from the AoIR Ethics Working Committee"; Markham and Buchanan, "Ethical Decision-Making and Internet Research."

# Bibliography

"♥ Why? ♥," 2023. https://fedipact.online/why.
Abbing, Roel Roscam, and Robert W. Gehl. "Shifting Your Research from X to Mastodon? Here's What You Need to Know." *Patterns* 5, no. 1 (January 12, 2024). https://doi.org/10.1016/j.patter.2023.100914.
AdaRoseCannon. "Support for Subscribing to Communal Block Lists · Issue #116 · Mastodon/Mastodon." *GitHub*, November 1, 2016. https://github.com/mastodon/mastodon/issues/116.
Akrich, Madeleine. "The De-Scription of Technical Objects." In *Shaping Technology/Building Society*, edited by Wiebe Bijker and John Law, 205–24. Cambridge, MA: MIT Press, 1992.
Allebach, J. P., and R. N. Stradling. "Computer-Aided Design of Dither Signals for Binary Display of Images." *Applied Optics* 18, no. 15 (August 1, 1979): 2708–13. https://doi.org/10.1364/AO.18.002708.
Allyn, Bobby. "Elon Musk Says NPR's 'State-Affiliated Media' Label Might Not Have Been Accurate." *NPR*, April 6, 2023, sec. Technology. https://www.npr.org/2023/04/06/1168455846/elon-musk-says-nprs-state-affiliated-media-label-might-not-have-been-accurate.
"Alphabet Announces Fourth Quarter and Fiscal Year 2022 Results." February 2, 2023. https://www.sec.gov/Archives/edgar/data/1652044/000165204423000013/googexhibit991q42022.htm.
"Apps API Methods—Mastodon Documentation." Accessed October 13, 2023. https://docs.joinmastodon.org/methods/apps/.
Arrington, Michael. "Details Revealed: Google OpenSocial to Launch Thursday." *TechCrunch*, October 31, 2007. http://www.techcrunch.com/2007/10/30/details-revealed-google-opensocial-to-be-common-apis-for-building-social-apps/.
ashe dryden [@ashedryden]. "I Will Not Attend or Speak at Conferences or Other Events That Do Not Have a Code of Conduct. #CoCPledge." Tweet. *Twitter*, February 3, 2014. https://twitter.com/ashedryden/status/430443215863291904.
Atton, Chris. *Alternative Media*. London: Sage, 2002.
Aubin, Christopher St, and Galen Stocking. "Key Facts About Gab." *Pew Research Center* (blog), January 24, 2023. https://www.pewresearch.org/short-reads/2023/01/24/key-facts-about-gab/.
Baker, J. Wayne. "The Covenantal Basis for the Development of Swiss Political Federalism: 1291–1848." *Publius: The Journal of Federalism* 23, no. 2 (January 1, 1993): 19–42. https://doi.org/10.1093/oxfordjournals.pubjof.a038067.
Baldwin, James. *The Fire Next Time*. New York: Vintage International, 1993.
Barratt, Monica J., and Alexia Maddox. "Active Engagement with Stigmatised Communities Through Digital Ethnography." *Qualitative Research*, May 22, 2016, 701–19. https://doi.org/10.1177/1468794116648766.
Barwick, Ryan. "X Ad Costs and Engagement Continue to Fall, per Report." Marketing Brew, March 27, 2024. https://www.marketingbrew.com/stories/2024/03/27/x-ad-costs-engagement-continue-to-fall-per-report.
Bass, Lisa. "When Care Trumps Justice: The Operationalization of Black Feminist Caring in Educational Leadership." *International Journal of Qualitative Studies in Education* 25, no. 1 (February 1, 2012): 73–87. https://doi.org/10.1080/09518398.2011.647721.
Baym, Nancy K. *Tune in, Log on: Soaps, Fandom and Online Community*. Thousand Oaks, CA: Sage, 2000.

BBC News. "Twitter in India: Why Was Rival Mastodon Trending?" November 8, 2019, sec. India. https://www.bbc.com/news/world-asia-india-50343054.

BBC News. "Twitter Insiders: We Can't Protect Users from Trolling Under Musk." March 5, 2023, sec. Technology. https://www.bbc.com/news/technology-64804007.

Berners-Lee, Tim. "W3C NOTE Process." June 1996. https://www.w3.org/Consortium/Process/NOTE.html.

Bielenberg, A., L. Helm, A. Gentilucci, D. Stefanescu, and Honggang Zhang. "The Growth of Diaspora—A Decentralized Online Social Network in the Wild." In *2012 IEEE Conference on Computer Communications Workshops (INFOCOM WKSHPS)*, 13–18, March, 2012. https://doi.org/10.1109/INFCOMW.2012.6193476.

Binswanger, Mathias. "Is There a Growth Imperative in Capitalist Economies? A Circular Flow Perspective." *Journal of Post Keynesian Economics* 31, no. 4 (2009): 707–27.

Bivens, Rena. "Under the Hood: The Software in Your Feminist Approach." *Feminist Media Studies* 15, no. 4 (June 19, 2015): 714–17. https://doi.org/10.1080/14680777.2015.1053717.

Blue, Violet. "'But He Does Good Work.'" *Medium*, June 16, 2016. https://medium.com/@violetblue/but-he-does-good-work-6710df9d9029#.2gz9toute.

Boellstorff, Tom. *Coming of Age in Second Life: An Anthropologist Explores the Virtually Human*. Princeton, NJ: Princeton University Press, 2008.

Booth, Robert. "Facebook Reveals News Feed Experiment to Control Emotions." *The Guardian*, June 29, 2014, sec. Technology. https://www.theguardian.com/technology/2014/jun/29/facebook-users-emotions-news-feeds.

Bowker, Geoffrey C., and Susan Leigh Star. *Sorting Things Out: Classification and Its Consequences*. Inside Technology. Cambridge, MA: MIT Press, 1999.

Brandon, John. "Instagram Is Outpacing TikTok for User Growth. Here's Why." *Forbes*, June 24, 2023. https://www.forbes.com/sites/johnbbrandon/2023/06/24/instagram-is-outpacing-tiktok-for-user-growth-heres-why/.

Braun, Josh. *This Program Is Brought to You by—: Distributing Television News Online*. New Haven, CT: Yale University Press, 2015.

Briant, Emma L. "Cambridge Analytica and SCL—How I Peered Inside the Propaganda Machine." *The Conversation*, April 17, 2018. http://theconversation.com/cambridge-analytica-and-scl-how-i-peered-inside-the-propaganda-machine-94867.

Briant, Emma L. "Researching Influence Operations: 'Dark Arts' Mercenaries and the Digital Influence Industry." In *The Oxford Handbook of Digital Diplomacy*, edited by Corneliu Bjola and Ilan Manor, 80–100. Oxford University Press, 2024. https://doi.org/10.1093/oxfordhb/9780192859198.013.5.

Brice, Sage. "Geographies of Vulnerability: Mapping Transindividual Geometries of Identity and Resistance." *Transactions of the Institute of British Geographers* 45, no. 3 (2020): 664–77. https://doi.org/10.1111/tran.12358.

Brinberg, Miriam, Nilam Ram, Jinping Wang, S. Shyam Sundar, James J. Cummings, Leo Yeykelis, and Byron Reeves. "Screenertia: Understanding 'Stickiness' of Media Through Temporal Changes in Screen Use." *Communication Research* 50, no. 5 (July 1, 2023): 535–60. https://doi.org/10.1177/00936502211062778.

Brock, André L. *Distributed Blackness: African American Cybercultures*. Critical Cultural Communication. New York: New York University Press, 2019.

Brodkin, Jon. "Musk Labels CBC '69% Government-Funded' as More News Outlets Quit Twitter." *Ars Technica*, April 18, 2023. https://arstechnica.com/tech-policy/2023/04/musk-feuds-with-another-public-broadcaster-labels-cbc-69-government-funded/.

Bucher, Taina. *If ... Then: Algorithmic Power and Politics*. Oxford Studies in Digital Politics. New York: Oxford University Press, 2018.

Bucher, Taina. "Want to Be on the Top? Algorithmic Power and the Threat of Invisibility on Facebook." *New Media & Society* 14, no. 7 (2012): 1164–80.

Burgess, Jean, and Ariadna Matamoros-Fernández. "Mapping Sociocultural Controversies Across Digital Media Platforms: One Week of #gamergate on Twitter, YouTube, and

Tumblr." *Communication Research and Practice* 2, no. 1 (January 2, 2016): 79–96. https://doi.org/10.1080/22041451.2016.1155338.

Cadwalladr, Carole, and Emma Graham-Harrison. "Revealed: 50 Million Facebook Profiles Harvested for Cambridge Analytica in Major Data Breach." *The Guardian*, March 17, 2018, sec. News. https://www.theguardian.com/news/2018/mar/17/cambridge-analytica-facebook-influence-us-election.

Caelin, Derek. "Decentralized Networks vs. the Trolls." In *Fundamental Challenges to Global Peace and Security: The Future of Humanity*, edited by Hoda Mahmoudi, Michael H. Allen, and Kate Seaman, 143–68. Cham: Springer, 2022. https://doi.org/10.1007/978-3-030-79072-1_8.

Captain, Sean. "Can Mastodon Be a Twitter Refuge for Marginalized Groups?" *Fast Company*, November 30, 2022. https://www.fastcompany.com/90817452/can-mastodon-be-a-twitter-refuge-for-marginalized-groups.

Card, Claudia. "Caring and Evil." Edited by Nel Noddings. *Hypatia* 5, no. 1 (1990): 101–8.

carrotcypher. "AMA with Eugen Rochko, Founder and Lead Developer of Mastodon, a Decentralized, Open-Source Social Media Platform Based on Open Web Protocols. Ask Your Questions Here!" Reddit Post. *R/Mastodon*, December 20, 2022. www.reddit.com/r/Mastodon/comments/zqfr4h/ama_with_eugen_rochko_founder_and_lead_developer/.

Chambers, Tim. "Project92 and the Fediverse—A Smarter Battle Plan to Protect the Open Social Web." June 23, 2023. https://www.timothychambers.net/2023/06/23/project-and-the.html.

Chan, Rosalie. "Open-Source Developers Are Burning Out, Quitting, and Even Sabotaging Their Own Projects—And It's Putting the Entire Internet at Risk." *Business Insider*, March 19, 2022. https://www.businessinsider.com/open-source-developers-burnout-low-pay-internet-2022-3.

chaos.social info. "Chaos.Social Info." Blog, November 2023. https://meta.chaos.social/money#report.

Chatterjee, Mohar. "Social Media's Fragmented Future." *Politico*, April 1, 2024. https://www.politico.com/newsletters/digital-future-daily/2023/06/22/social-medias-fragmented-future-00103238.

Cherry, Myisha V. *The Case for Rage: Why Anger Is Essential to Anti-Racist Struggle*. New York: Oxford University Press, 2021.

Chess, Shira, and Adrienne Shaw. "A Conspiracy of Fishes, or, How We Learned to Stop Worrying About #GamerGate and Embrace Hegemonic Masculinity." *Journal of Broadcasting & Electronic Media* 59, no. 1 (January 2, 2015): 208–20. https://doi.org/10.1080/08838151.2014.999917.

Choudhury, Deepsekhar, and Vikas Sn. "Exclusive: Meta Mulls a Twitter Competitor Codenamed 'P92' That Will Be Interoperable with Mastodon." *Moneycontrol*, March 10, 2023. https://www.moneycontrol.com/news/business/startup/meta-mulls-a-twitter-competitor-codenamed-p92-that-will-be-interoperable-with-mastodon-10223961.html.

Claburn, Thomas. "Meta, Which Pays for Web Scraping, Sues to Stop Web Scraping." *The Register*, February 2, 2023. https://www.theregister.com/2023/02/02/meta_web_scraping/.

Clouse, Jay. "Threads: My Thoughts So Far." *Creator Science*, July 9, 2023. https://creatorscience.com/threads-launch/.

Cohen, Noam. "After Years of Abusive E-Mails, the Creator of Linux Steps Aside." *The New Yorker*, September 19, 2018. https://www.newyorker.com/science/elements/after-years-of-abusive-e-mails-the-creator-of-linux-steps-aside.

Cole, Samantha. "Switter, the Twitter for Sex Workers, Is Shutting Down." *Vice* (blog), February 14, 2022. https://www.vice.com/en/article/7kb7vx/switter-the-twitter-for-sex-workers-is-shutting-down.

Coleman, E. Gabriella. *Coding Freedom: The Ethics and Aesthetics of Hacking*. Princeton, NJ: Princeton University Press, 2013.

Coleman, E. Gabriella. *Hacker, Hoaxer, Whistleblower, Spy: The Many Faces of Anonymous.* New York: Verso, 2014.

Conger, Kate, and Lauren Hirsch. "Elon Musk Completes $44 Billion Deal to Own Twitter." *The New York Times*, October 28, 2022, sec. Technology. https://www.nytimes.com/2022/10/27/technology/elon-musk-twitter-deal-complete.html.

"Contributor Covenant: A Code of Conduct for Open Source and Other Digital Commons Communities." 2014. https://www.contributor-covenant.org/.

corbet. "The Open Collective Foundation Is Shutting down [LWN.Net]." February 28, 2024. https://lwn.net/Articles/963958/.

Couldry, Nick. *Media Rituals: A Critical Approach.* London: Routledge, 2003.

Couldry, Nick, and James Curran, eds. *Contesting Media Power: Alternative Media in a Networked World.* Oxford: Rowman & Littlefield, 2003.

Cuthbertson, Anthony. "Project Ghostbusters: Facebook Spied on Snapchat Users, Documents Reveal." *The Independent*, March 27, 2024, sec. Tech. https://www.independent.co.uk/tech/facebook-spy-snapchat-zuckerberg-meta-b2519458.html.

Dame-Griff, Avery. *The Two Revolutions: A History of the Transgender Internet.* Queer/Trans/Digital. New York: New York University Press, 2023.

D'Angelo, Adam. "What Specific Actions in the Early Years (2004–2007) Led to the Massive Rift Between Facebook and Google? Is Anything Related to Sheryl S." *Quora*, 2011. https://www.quora.com/What-specific-actions-in-the-early-years-2004-2007-led-to-the-massive-rift-between-Facebook-and-Google-Is-anything-related-to-Sheryl-Sandbergs-movement-to-Facebook/answer/Adam-DAngelo.

Davis, Wes. "Threads Is Officially Starting to Test ActivityPub Integration." *The Verge*, December 13, 2023. https://www.theverge.com/2023/12/13/24000120/threads-meta-activitypub-test-mastodon.

De Schutter, Helder. "Federalism as Fairness." *Journal of Political Philosophy* 19, no. 2 (June 2011): 167–89. https://doi.org/10.1111/j.1467-9760.2010.00368.x.

Decker, Kris De. "About the Solar Powered Website." *LOW←TECH MAGAZINE*, April 22, 2022. https://solar.lowtechmagazine.com/about/the-solar-website/.

Dehghan, Ehsan, and Ashwin Nagappa. "Politicization and Radicalization of Discourses in the Alt-Tech Ecosystem: A Case Study on Gab Social." *Social Media + Society* 8, no. 3 (July 1, 2022): 20563051221113075. https://doi.org/10.1177/20563051221113075.

Deibert, Ronald J. "The Road to Digital Unfreedom: Three Painful Truths About Social Media." *Journal of Democracy* 30, no. 1 (January 9, 2019): 25–39. https://doi.org/10.1353/jod.2019.0002.

Delfanti, Alessandro. "Machinic Dispossession and Augmented Despotism: Digital Work in an Amazon Warehouse." *New Media & Society* 23, no. 1 (January 1, 2021): 39–55. https://doi.org/10.1177/1461444819891613.

"Differences in Mastodon API Responses from Vanilla Mastodon—Pleroma Documentation." Accessed October 13, 2023. https://docs-develop.pleroma.social/backend/development/API/differences_in_mastoapi_responses/.

diogo. "Diogo/Gnu-Social," June 11, 2019. https://notabug.org/diogo/gnu-social/src/master/CODE_OF_CONDUCT.md.

DiResta, Renee, Kris Shaffer, Becky Ruppel, David Sullivan, Robert Matney, Ryan Fox, Jonathan Albright, Ben Johnson, and Canfield Research. "The Tactics & Tropes of the Internet Research Agency." *New Knowledge*, 2018. https://www.intelligence.senate.gov/sites/default/files/documents/NewKnowledge-Disinformation-Report-Whitepaper.pdf.

Doctorow, Cory. "Unpersoned." *Pluralistic* (blog), July 22, 2024. https://pluralistic.net/2024/07/22/degoogled/.

Driscoll, Kevin. *The Modem World: A Prehistory of Social Media.* New Haven, CT: Yale University Press, 2022.

Du Mez, Kristin Kobes. *Jesus and John Wayne: How White Evangelicals Corrupted a Faith and Fractured a Nation.* 1st ed. New York: Liveright, 2020.

Duffy, Brooke Erin. "The Romance of Work: Gender and Aspirational Labour in the Digital Culture Industries." *International Journal of Cultural Studies* 19, no. 4 (July 1, 2016): 441–57. https://doi.org/10.1177/1367877915572186.

Duffy, Brooke Erin, Thomas Poell, and David B. Nieborg. "Platform Practices in the Cultural Industries: Creativity, Labor, and Citizenship." *Social Media + Society* 5, no. 4 (October 1, 2019): 2056305119879672. https://doi.org/10.1177/2056305119879672.

Duguay, Stefanie, Jean Burgess, and Nicolas Suzor. "Queer Women's Experiences of Patchwork Platform Governance on Tinder, Instagram, and Vine." *Convergence*, June 19, 2018, 1354856518781530. https://doi.org/10.1177/1354856518781530.

Duguay, Stefanie, Jean Burgess, and Nicolas Suzor. "Queer Women's Experiences of Patchwork Platform Governance on Tinder, Instagram, and Vine." *Convergence* 26, no. 2 (April 1, 2020): 237–52. https://doi.org/10.1177/1354856518781530.

Dunbar-Hester, Christina. *Hacking Diversity*. Princeton, NJ: Princeton University Press, 2019. https://press.princeton.edu/books/paperback/9780691192888/hacking-diversity.

Dunbar-Hester, Christina. "Showing Your Ass on Mastodon: Lossy Distribution, Hashtag Activism, and Public Scrutiny on Federated, Feral Social Media." *First Monday*, March 9, 2024. https://doi.org/10.5210/fm.v29i3.13367.

Dyer-Witheford, Nick. *Cyber-Marx: Cycles and Circuits of Struggle in High-Technology Capitalism*. Urbana: University of Illinois Press, 1999.

Ehmke, Coraline Ada. "Codes of Conduct: When Being Excellent Is Not Enough." Blog. *Model View Culture* (blog), December 10, 2014. https://modelviewculture.com/pieces/codes-of-conduct-when-being-excellent-is-not-enough.

Elazar, Daniel J. "Confederation and Federal Liberty." *Publius* 12, no. 4 (1982): 1–14. https://doi.org/10.2307/3329660.

Elazar, Daniel J. "Covenant." In *Encyclopedia of Politics and Religion*, 2nd ed., 217–21. Washington. DC: CQ Press, 2007. https://doi.org/10.4135/9781483300535.

Elliott, Karla. "Caring Masculinities: Theorizing an Emerging Concept." *Men and Masculinities* 19, no. 3 (August 1, 2016): 240–59. https://doi.org/10.1177/1097184X15576203.

Elon Musk [@elonmusk]. "Free Speech Is Essential to a Functioning Democracy. Do You Believe Twitter Rigorously Adheres to This Principle?" Tweet. *Twitter*, March 25, 2022. https://twitter.com/elonmusk/status/1507259709224632344.

Elon Musk [@elonmusk]. "The Bird Is Freed." Tweet. *Twitter*, October 28, 2022. https://twitter.com/elonmusk/status/1585841080431321088.

Emerson, Sarah. "As Literal Nazis Abuse Twitter, CEO Asks How to Improve the Platform." *Vice* (blog), December 30, 2016. https://www.vice.com/en/article/kb79ew/as-literal-nazis-abuse-twitter-ceo-asks-how-to-improve-the-platform-dorsey.

Emsley, Iain. "Exploring Mastodon: A Proof of Concept Tool." Mastodon Research Symposium University of Warwick, June 6, 2023.

Enyedy, Edgar, and Nathaniel Tkacz. "'Good Luck with Your WikiPAIDia': Reflections on the 2002 Fork of the Spanish Wikipedia." Blog. Network Cultures, January 2011. http://networkcultures.org/wpmu/cpov/lang/de/2011/01/15/spanish_fork/.

Ernst, Johannes. "Congratulations!" December 15, 2023. https://lists.w3.org/Archives/Public/public-swicg/2023Dec/0058.html.

Ess, Charles, and AoIR Ethics Working Committee. "Ethical Decision-Making and Internet Research: Recommendations from the Aoir Ethics Working Committee." Association for Internet Researchers, November 7, 2002. www.aoir.org/reports/ethics.pdf.

"The Facebook Files, Part 4: The Outrage Algorithm." *The Journal*. Accessed July 1, 2024. https://www.wsj.com/podcasts/the-journal/the-facebook-files-part-4-the-outrage-algorithm/e619fbb7-43b0-485b-877f-18a98ffa773f.

Feathers, Todd, Simon Fondrie-Teitler, Angie Waller, and Surya Mattu. "Facebook Is Receiving Sensitive Medical Information from Hospital Websites—The Markup." *The Markup*, June 16, 2022. https://themarkup.org/pixel-hunt/2022/06/16/facebook-is-receiving-sensitive-medical-information-from-hospital-websites.

FediBlock. "FediBlock," January 10, 2021. https://web.archive.org/web/20210110104353/http://fediblock.org/.
"#Fedipact—The Instances Blocking Zuckerberg's Threads.Net." 2023. https://fedipact.veganism.social/?v2.
*Feels Good Man*. Documentary, Comedy. Ready Fictions, Wavelength, XTR, 2020.
Finley, Klint. "The Woman Bringing Civility to Open Source Projects." *Wired*, September 26, 2018. https://www.wired.com/story/woman-bringing-civility-to-open-source-projects/.
Fisher, Berenice, and Joan Tronto. "Towards a Feminist Theory of Caring." In *Circles of Care: Work and Identity in Women's Lives*, edited by Emily K. Abel and Margaret K. Nelson, 35–62. SUNY Series on Women and Work. Albany: State University of New York Press, 1990.
Fisher, Mark. *Capitalist Realism: Is There No Alternative?* Winchester, UK: Zero Books, 2009.
Flowers, Johnathan. "The Whiteness of Mastodon." Interview by Justin Hendrix. Audio. November 23, 2022. https://techpolicy.press/the-whiteness-of-mastodon/.
Flynn, Adam. "Solarpunk: Notes Toward a Manifesto." Blog. Project Hieroglyph, September 4, 2014. https://hieroglyph.asu.edu/2014/09/solarpunk-notes-toward-a-manifesto/.
Fopp, Rodney. "'Repressive Tolerance': Herbert Marcuse's Exercise in Social Epistemology." *Social Epistemology* 24, no. 2 (April 1, 2010): 105–22. https://doi.org/10.1080/02691721003749901.
Ford, Winston "Stone." "On Threads, SPILL, and Social Media Being 'Cool' Again." *Medium* (blog), July 6, 2023. https://medium.com/@thisisstone/on-threads-spill-and-social-media-being-cool-again-92bbb6bb6a82.
fpiesche. "[Feature] Customisable Code of Conduct/Terms of Service Page · Issue #411 · Mastodon/Mastodon." *GitHub*, January 5, 2017. https://github.com/mastodon/mastodon/issues/411.
Free Software Foundation. "What Is Free Software?" December 20, 2013. https://www.gnu.org/philosophy/free-sw.html.
Frenkel, Sheera. "QAnon Is Still Spreading on Facebook, Despite a Ban." *The New York Times*, December 18, 2020, sec. Technology. https://www.nytimes.com/2020/12/18/technology/qanon-is-still-spreading-on-facebook-despite-a-ban.html.
Frenkel, Sheera, and Kate Conger. "Hate Speech's Rise on Twitter Is Unprecedented, Researchers Find." *The New York Times*, December 2, 2022, sec. Technology. https://www.nytimes.com/2022/12/02/technology/twitter-hate-speech.html.
Frenkel, Sheera, and Stuart A. Thompson. "Some Far-Right Accounts on Twitter Saw Surge in New Followers, Researchers Say." *The New York Times*, October 28, 2022, sec. Technology. https://www.nytimes.com/2022/10/28/technology/twitter-far-right-followers.html.
Freitas, Miguel. "Twister—A P2P microblogging platform." arXiv, 2013. http://arxiv.org/abs/1312.7152.
Fuchs, Christian. "A Contribution to the Critique of the Political Economy of Transnational Informational Capitalism." *Rethinking Marxism: A Journal of Economics, Culture & Society* 21, no. 3 (2009): 387. https://doi.org/10.1080/08935690902955104.
Fuchs, Christian. *Critical Theory of Communication*. Westminster: University of Westminster Press, 2016. http://www.uwestminsterpress.co.uk/site/books/detail/1/critical-theory-of-communication/.
Fuchs, Christian. "The Political Economy of Privacy on Facebook." *Television & New Media* 13, no. 2 (March 1, 2012): 139–59. https://doi.org/10.1177/1527476411415699.
Fuller, Matthew. *Software Studies: A Lexicon*. Cambridge, MA: MIT Press, 2008.
Fung, Brian. "Elon Musk's Twitter Blocked Links to Rival Mastodon. That Could Raise Alarms Among Regulators." CNN, December 16, 2022. https://www.cnn.com/2022/12/16/tech/mastodon-twitter-links/index.html.
"Further Supplementary Written Evidence on Cambridge Analytica, Leave.EU and Eldon Insurance." London, October 2018. http://data.parliament.uk/writtenevidence/committeeevidence.svc/evidencedocument/digital-culture-media-and-sport-committee/disinformation-and-fake-news/written/92380.html.

"Gab Employee Explains Why ActivityPub Sucks." 2021. https://www.youtube.com/watch?v=3kDtZ8MBWy8.

Gab Social. "Rob Colbert on Gab: 'Federation. Would You Miss It?'" May 23, 2020. https://gab.com/shadowknight412/posts/104219677077942121.

"Gab (Social Network)." In *Wikipedia*, March 13, 2024. https://en.wikipedia.org/w/index.php?title=Gab_(social_network)&oldid=1213455179.

Gallagher, Erin. "Advanced Meme Warfare—July 6, 2016 (CW)." *Medium*, October 1, 2018. https://medium.com/@erin_gallagher/advanced-meme-warfare-july-6-2016-cw-5f9287ef36cd.

Gallaher, Carolyn. "Mainstreaming White Supremacy: A Twitter Analysis of the American 'Alt-Right.'" *Gender, Place & Culture* 28, no. 2 (February 1, 2021): 224–52. https://doi.org/10.1080/0966369X.2019.1710472.

Gargron. "Contributor Covenant Code of Conduct." Ruby. 2016. Reprint, Mastodon, November 13, 2017. https://github.com/mastodon/mastodon/blob/1122697b375da82cbb156b73eb1015ae066fc6ee/CODE_OF_CONDUCT.md.

Gehl, Robert W. "Alternative Social Media: From Critique to Code." In *The SAGE Handbook of Social Media*, edited by Jean Burgess, Alice E. Marwick, and Thomas Poell, 330–52. Thousand Oaks, CA: Sage, 2017.

Gehl, Robert W. "Building a Better Twitter: A Study of the Twitter Alternatives GNU Social, Quitter, Rstat.Us, and Twister." *The Fibreculture Journal*, no. 26 (December 22, 2015): 60–86. https://doi.org/10.15307/fcj.26.190.2015.

Gehl, Robert W. "The Case for Alternative Social Media." *Social Media + Society* 1, no. 2 (July 1, 2015): 2056305115604338. https://doi.org/10.1177/2056305115604338.

Gehl, Robert W. "Critical Reverse Engineering: The Case of Twitter and TalkOpen." In *Compromised Data: From Social Media to Big Data*, edited by Ganaele Langlois, Joanna Redden, and Greg Elmer, 147–69. New York: Bloomsbury, 2015.

Gehl, Robert W. "Elon Musk's Stance on Free Speech Doesn't Include Competition to Twitter." *Toronto Star*, December 18, 2022, sec. Opinion. https://www.thestar.com/opinion/contributors/2022/12/18/elon-musks-stance-on-free-speech-doesnt-include-competition-to-twitter.html.

Gehl, Robert W. "Power/Freedom on the Dark Web: A Digital Ethnography of the Dark Web Social Network." *New Media and Society*, October 16, 2014, 1–17.

Gehl, Robert W. *Reverse Engineering Social Media: Software, Culture, and Political Economy in New Media Capitalism*. Philadelphia, PA: Temple University Press, 2014.

Gehl, Robert W. "The Politics of Punctualization and Depunctualization in the Digital Advertising Alliance." *The Communication Review* 19, no. 1 (January 2, 2016): 35–54. https://doi.org/10.1080/10714421.2016.1128187.

Gehl, Robert W. *Weaving the Dark Web: Legitimacy on Freenet, Tor, and I2P*. Cambridge, MA: MIT Press, 2018.

Gehl, Robert W. "'Why I Left Facebook': Stubbornly Refusing to Not Exist Even After Opting Out of Mark Zuckerberg's Social Graph." In *Unlike Us Reader: Social Media Monopolies and Their Alternatives*, edited by Geert Lovink and Miriam Rausch, 220–38. Amsterdam: Institute of Network Cultures, 2013.

Gehl, Robert W., and Sean T. Lawson. *Social Engineering: How Crowdmasters, Phreaks, Hackers, and Trolls Create a New Form of Manipulative Communication*. Cambridge, MA: MIT Press, 2022.

Gehl, Robert W., and Julie Snyder-Yuly. "The Need for Social Media Alternatives." *Democratic Communiqué* 27, no. 1 (August 2, 2016): 78.

Gehl, Robert W., and Diana Zulli. "The Digital Covenant: Non-Centralized Platform Governance on the Mastodon Social Network." *Information, Communication & Society* 26, no. 16 (December 15, 2022): 3275–91. https://doi.org/10.1080/1369118X.2022.2147400.

Gerken, Tom. "Twitter Blocks Users from Sharing Mastodon Links." BBC News, December 16, 2022. https://www.bbc.com/news/technology-63999452.

ghost. "Custom Federation Levels (at the Very Least, for Private Posts) · Issue #712 · Mastodon/ Mastodon." *GitHub*, April 1, 2017. https://github.com/mastodon/mastodon/issues/712.

Gibron, PopMatters. "Depth of Field: Auteur Theory—Godmonster of Indian Flats," September 27, 2006. https://www.popmatters.com/depth-of-field-auteur-theory-godmonster-of-indian-flats-2495711223.html.

Gibson, Anna. "Free Speech and Safe Spaces: How Moderation Policies Shape Online Discussion Spaces." *Social Media + Society* 5, no. 1 (January 2019): 205630511983258. https://doi.org/10.1177/2056305119832588.

Gibson-Graham, J. K. "Diverse Economies: Performative Practices for 'Other Worlds.'" *Progress in Human Geography* 32, no. 5 (October 1, 2008): 613–32. https://doi.org/10.1177/0309132508090821.

Gillespie, Tarleton. *Custodians of the Internet: Platforms, Content Moderation, and the Hidden Decisions That Shape Social Media*. New Haven, CT: Yale University Press, 2018.

Gilligan, Carol. *In a Different Voice: Psychological Theory and Women's Development*. Cambridge, MA: Harvard University Press, 1982.

Gleason, Alex. "Soapbox BE Is Now Rebased." August 19, 2022. https://soapbox.pub/blog/soapbox-be-is-now-rebased/.

Goldstein, Matthew, and Ryan Mac. "Trump's Truth Social Is Poised to Join a Crowded Field." *The New York Times*, February 18, 2022, sec. Business. https://www.nytimes.com/2022/02/18/business/trumps-truth-social.html.

Goodwin, Jazmin. "Gab: Everything You Need to Know About the Fast-Growing, Controversial Social Network." CNN, January 17, 2021. https://www.cnn.com/2021/01/17/tech/what-is-gab-explainer/index.html.

Gordon, Myron J., and Jeffrey S. Rosenthal. "Capitalism's Growth Imperative." *Cambridge Journal of Economics* 27, no. 1 (January 1, 2003): 25–48. https://doi.org/10.1093/cje/27.1.25.

Gorur, Radhika, Mary Hamilton, Christian Lundahl, and Elin Sundström Sjödin. "Politics by Other Means? STS and Research in Education." *Discourse: Studies in the Cultural Politics of Education* 40, no. 1 (January 2, 2019): 1–15. https://doi.org/10.1080/01596306.2018.1549700.

Gorwa, Robert. "What Is Platform Governance?" *Information, Communication & Society* 22, no. 6 (May 12, 2019): 854–71. https://doi.org/10.1080/1369118X.2019.1573914.

Goujard, Clothilde. "EU Charges Elon Musk's X for Letting Disinfo Run Wild." *Politico*, July 12, 2024. https://www.politico.eu/article/eu-charges-musks-x-for-letting-disinfo-run-wild/.

Gow, Gordon A. "Alternative Social Media for Outreach and Engagement: Considering Technology Stewardship as a Pathway to Adoption." *ERA*, January 1, 2020. https://doi.org/10.7939/r3-xy8j-w962.

Greenwood, Faine. "How I Accidentally Ruined Bluesky with Pictures of Sexy Alf." Substack newsletter. *Little Flying Robots* (blog), May 5, 2023. https://faineg.substack.com/p/how-i-accidentally-ruined-bluesky.

Grossman, Lev. "Mark Zuckerberg—Person of the Year 2010." *Time*, December 15, 2010. http://www.time.com/time/specials/packages/article/0,28804,2036683_2037183_2037185,00.html.

Gruber, Jon. "More on Preemptively Blocking Facebook's Imminent ActivityPub Entry." *Daring Fireball*, June 20, 2023. https://daringfireball.net/2023/06/more_on_preemptively_blocking.

Guy, Amy. "The Presentation of Self on a Decentralised Web." University of Edinburgh, 2017. http://dr.amy.gy/.

Hallinan, Blake, and Ted Striphas. "Recommended for You: The Netflix Prize and the Production of Algorithmic Culture." *New Media & Society* , June 23, 2014, 1461444814538646. https://doi.org/10.1177/1461444814538646.

Hampton, Rachelle. "The Black Feminists Who Saw the Alt-Right Threat Coming." *Slate Magazine*, April 23, 2019. https://slate.com/technology/2019/04/black-feminists-alt-right-twitter-gamergate.html.

Han, Catherine, Deepak Kumar, and Zakir Durumeric. "On the Infrastructure Providers That Support Misinformation Websites." *Proceedings of the International AAAI Conference on Web and Social Media* 16 (May 31, 2022): 287–98. https://doi.org/10.1609/icwsm.v16i1.19292.

Hanhardt, Christina B. *Safe Space: Gay Neighborhood History and the Politics of Violence*. Durham, NC: Duke University Press, 2013.

hannah. "Akkoma: A Vision to Refocus Pleroma." *Coffee and Dreams*, June 24, 2022. https://coffee-and-dreams.uk.

Hardcastle, Jessica Lyons. "Twitter Admits 'Security Incident' Broke Circle Privacy." May 8, 2023. https://www.theregister.com/2023/05/08/twitter_circle_security_incident/.

Hart, ALLIE ❤. "Mourning Mastodon." *Medium* (blog), April 23, 2017. https://medium.com/@alliethehart/gameingers-are-dead-and-so-is-mastodon-705b535ed616.

Hassler-Forest, Dan. "'When You Get There, You Will Already Be There' Stranger Things, Twin Peaks and the Nostalgia Industry." *Science Fiction Film and Television* 13, no. 2 (2020): 175–97.

Heath, Alex. "This Is What Instagram's Upcoming Twitter Competitor Looks Like." *The Verge*, June 8, 2023. https://www.theverge.com/2023/6/8/23754304/instagram-meta-twitter-competitor-threads-activitypub.

Henry, John. "Mastodon Is Dead in the Water." *Hacker Noon*, April 5, 2017. https://hackernoon.com/mastodon-is-dead-in-the-water-888c10e8abb1.

Heydari, Anis. "Apple Looks for More Money by Changing the Math on Patreon." CBC News, August 16, 2024. https://www.cbc.ca/news/business/apple-patreon-app-store-1.7296083.

Hickel, Jason, Giorgos Kallis, Tim Jackson, Daniel W. O'Neill, Juliet B. Schor, Julia K. Steinberger, Peter A. Victor, and Diana Ürge-Vorsatz. "Degrowth Can Work—Here's How Science Can Help." *Nature* 612, no. 7940 (December 2022): 400–403. https://doi.org/10.1038/d41586-022-04412-x.

Hickey, Daniel, Matheus Schmitz, Daniel Fessler, Paul Smaldino, Goran Muric, and Keith Burghardt. "Auditing Elon Musk's Impact on Hate Speech and Bots." *arXiv*, April 13, 2023. http://arxiv.org/abs/2304.04129.

Hightower, Ray. "An Environment of Respect." Blog. rayhightower.com, January 21, 2014. https://rayhightower.com/blog/2014/01/21/an-environment-of-respect/.

Hogan, Mél. "Data Flows and Water Woes: The Utah Data Center." *Big Data & Society* 2, no. 2 (July 1, 2015): 2053951715592429. https://doi.org/10.1177/2053951715592429.

Holt, Kristoffer. *Right-Wing Alternative Media*. London: Routledge, 2020.

Hoover, Amanda. "The Mastodon Bump Is Now a Slump." *Wired*, February 7, 2023. https://www.wired.com/story/the-mastodon-bump-is-now-a-slump/.

Hoover, Amanda. "Twitter Users Have Caused a Mastodon Meltdown." *Wired*, November 9, 2022. https://www.wired.com/story/twitter-users-mastodon-meltdown/.

Horowitz, Irving Louis. "Daniel J. Elazar and the Covenant Tradition in Politics." *Publius: The Journal of Federalism* 31, no. 1 (January 1, 2001): 1–7. https://doi.org/10.1093/oxfordjournals.pubjof.a004878.

Howard, Philip N, Bharath Ganesh, Dimitra Liotsiou, John Kelly, and Camille François. "The IRA, Social Media and Political Polarization in the United States, 2012–2018." Oxford: Computational Propaganda Research Project, 2018.

I ask questions. "List of Independent GNU Social Instances." Wiki, February 20, 2017. https://web.archive.org/web/20170220220116/http://skilledtests.com/wiki/List_of_Independent_GNU_social_Instances.

"I Believe They Broke. . . ." Reddit Comment. *R/Mastodon*, August 22, 2020. www.reddit.com/r/Mastodon/comments/ie49qp/deleted_by_user/g2g5ghs/.

Jasser, Greta, Jordan McSwiney, Ed Pertwee, and Savvas Zannettou. "'Welcome to #GabFam': Far-Right Virtual Community on Gab." *New Media & Society*, June 28, 2021, 14614448211024546. https://doi.org/10.1177/14614448211024546.

Jeong, Sarah. "Mastodon Is Like Twitter Without Nazis, So Why Are We Not Using It?" *Motherboard*, April 4, 2017. https://motherboard.vice.com/en_us/article/783akg/mastodon-is-like-twitter-without-nazis-so-why-are-we-not-using-it.

Johnston, Ian. "Twitter Rival Mastodon Rejects Funding to Preserve Nonprofit Status." *Ars Technica*, December 28, 2022. https://arstechnica.com/tech-policy/2022/12/twitter-rival-mastodon-rejects-funding-to-preserve-nonprofit-status/.

Join Mastodon. "Mastodon Annual Report 2022." 2023. https://joinmastodon.org/reports/Mastodon%20Annual%20Report%202022.pdf.

Jones, Jada, and Maria Diaz. "5 Things to Know About Meta's Threads App Before You Entangle Your Instagram Account." ZDNET, July 6, 2023. https://www.zdnet.com/article/5-things-to-know-about-metas-threads-app-before-you-entangle-your-instagram-account/.

Kaiser, Brittany. *Targeted: The Cambridge Analytica Whistleblower's Inside Story of How Big Data, Trump, and Facebook Broke Democracy and How It Can Happen Again.* Illustrated ed.. New York: Harper, 2019.

Kalbag, Laura. *Accessibility for Everyone.* New York: A Book Apart, 2017.

Kang, Jay Caspian. "What Elon Musk Doesn't Know About Free Speech." *The New Yorker*, December 6, 2022. https://www.newyorker.com/news/our-columnists/what-elon-musk-doesnt-know-about-free-speech.

Karpf, Dave. "Bluesky Is Just Twitter, Without the Burden of Everything Ruining Twitter." Substack newsletter. *The Future, Now and Then* (blog), May 18, 2023. https://davekarpf.substack.com/p/bluesky-is-just-twitter-without-the.

Katz, Michael L., and Carl Shapiro. "Systems Competition and Network Effects." *Journal of Economic Perspectives* 8, no. 2 (June 1994): 93–115. https://doi.org/10.1257/jep.8.2.93.

Kazemi, Darius. "How to Run a Small Social Network Site for Your Friends." August 31, 2019. https://runyourown.social/.

Kazemi, Darius. "A Highly Opinionated Guide to Learning About ActivityPub." *Tiny Subversions*, June 1, 2019. https://tinysubversions.com/notes/reading-activitypub/.

Kerr, Dara. "Twitter Once Muzzled Russian and Chinese State Propaganda. That's over Now." *NPR*, April 21, 2023, sec. Business. https://www.npr.org/2023/04/21/1171193551/twitter-once-muzzled-russian-and-chinese-state-propaganda-thats-over-now.

Kissane, Erin. "Untangling Threads." Erin Kissane's small internet website, December 21, 2023. https://erinkissane.com/untangling-threads.

Kissane, Erin, and Darius Kazemi. "Findings Report: Governance on Fediverse Microblogging Servers." *Manubot*, August 20, 2024. https://fediverse-governance.github.io/.

Kittay, Eva Feder. "The Ethics of Care, Dependence, and Disability." *Ratio Juris* 24, no. 1 (March 2011): 49–58. https://doi.org/10.1111/j.1467-9337.2010.00473.x.

Klein, Naomi. "Toxic Nostalgia, From Putin to Trump to the Trucker Convoys." *The Intercept*, March 1, 2022. https://theintercept.com/2022/03/01/war-climate-crisis-putin-trump-oil-gas/.

Klemens, Ben. "Mastodon—And the Pros and Cons of Moving beyond Big Tech Gatekeepers." *Ars Technica*, January 2, 2023. https://arstechnica.com/gadgets/2023/01/mastodon-highlights-pros-and-cons-of-moving-beyond-big-tech-gatekeepers/.

Knibbs, Kate. "I Regret to Inform You That Bluesky Is Fun." *Wired*, May 2, 2023. https://www.wired.com/story/bluesky-is-fun/.

Knight, Robb. "Meta Doesn't Need ActivityPub to Slurp Up Your Data." *Robb Knight*, December 15, 2023. https://rknight.me/blog/meta-doesnt-need-activitypub-to-slurp-up-your-data/.

Know Your Meme. "Stop Trying to Make Fetch Happen." June 4, 2014. https://knowyourmeme.com/memes/stop-trying-to-make-fetch-happen.

Kramer, Adam D. I., Jamie E. Guillory, and Jeffrey T. Hancock. "Experimental Evidence of Massive-Scale Emotional Contagion Through Social Networks." *Proceedings of the National Academy of Sciences*, June 2, 2014. https://doi.org/10.1073/pnas.1320040111.

Kuhn, Bradley M. "Trump's Social Media Platform and the Affero General Public License (of Mastodon)." Software Freedom Conservancy, October 21, 2021. https://sfconservancy.org/blog/2021/oct/21/trump-group-agplv3/.

Ladner, Kiera. "Treaty Federalism: An Indigenous Vision of Canadian Federalisms." In *New Trends in Canadian Federalism*, edited by Francois Rocher and Miriam Smith, 167–94. Peterborough, Ont.: Broadview Press, 2003.

Ladner, Kiera L. "Colonialism Isn't the Only Answer: Indigenous Peoples and Multilevel Governance in Canada." In *Federalism, Feminism and Multilevel Governance*, 1st ed., edited by Melissa Haussman, Marian Sawer, and Jill Vickers, 67–82. New York: Routledge, 2016. https://doi.org/10.4324/9781315582085-5.

Laser, Stefan, Anne Pasek, Ester Sørensen, Mél Hogan, Mace Ojala, Jens Fehrenbacher, Maximilian Gregor Hepach, Leman øelik, and Koushik Ravi Kumar. "The Environmental Footprint of Social Media Hosting: Tinkering with Mastodon." *EASST*, no. 3 (2022). https://www.easst.net/article/the-environmental-footprint-of-social-media-hosting-tinkering-with-mastodon/.

Lekach, Sasha. "The Coder Who Built Mastodon Is 24, Fiercely Independent, and Doesn't Care About Money." *Mashable*, April 7, 2017. https://mashable.com/2017/04/06/eugen-rochko-mastodon-interview/.

Lemmer-Webber, Christine. "Split off ActivityPub CG or WG." October 1, 2023.

Lemmer-Webber, Christine, Evan Henshaw-Plath, and Golda Vez. "Open Social Media: Origin Stories." University of Colorado–Boulder, June 13, 2023. https://archive.org/details/opensocialmedia-originstories.

Lemmer-Webber, Christine, Jessica Tallon, Erin Shephard, Amy Guy, and Evan Prodromou. "ActivityPub," January 23, 2018. https://www.w3.org/TR/activitypub/.

Lemmer-Webber, Morgan, and Christine Lemmer-Webber. "Fediverse Reflections While the Bird Burns." *FOSS and Crafts*, December 1, 2022. https://fossandcrafts.org/episodes/053-fediverse-reflections-while-the-bird-burns.html.

Li, Pengfei, Jianyi Yang, Mohammad A. Islam, and Shaolei Ren. "Making AI Less 'Thirsty': Uncovering and Addressing the Secret Water Footprint of AI Models." *arXiv*, April 6, 2023. https://doi.org/10.48550/arXiv.2304.03271.

Light, Ben, Jean Burgess, and Stefanie Duguay. "The Walkthrough Method: An Approach to the Study of Apps." *New Media & Society*, November 11, 2016. https://doi.org/10.1177/1461444816675438.

Mac, Ryan, and Kate Conger. "X May Lose Up to $75 Million in Revenue as More Advertisers Pull Out." *The New York Times*, November 24, 2023, sec. Business. https://www.nytimes.com/2023/11/24/business/x-elon-musk-advertisers.html.

Mac, Ryan, and Sheera Frenkel. "Internal Alarm, Public Shrugs: Facebook's Employees Dissect Its Election Role." *The New York Times*, October 22, 2021, sec. Technology. https://www.nytimes.com/2021/10/22/technology/facebook-election-misinformation.html.

Maher, Stephen. "Facebook's Algorithm Comes Under Scrutiny." Centre for International Governance Innovation, October 8, 2021. https://www.cigionline.org/articles/facebooks-algorithm-comes-under-scrutiny/.

Makuch, Ben. "The Nazi-Free Alternative to Twitter Is Now Home to the Biggest Far Right Social Network." *Vice* (blog), July 11, 2019. https://www.vice.com/en_ca/article/mb8y3x/the-nazi-free-alternative-to-twitter-is-now-home-to-the-biggest-far-right-social-network.

Mansoux, Aymeric, and Roel Roscam Abbing. "Seven Theses on the Fediverse and the Becoming of FLOSS." In *The Eternal Network*, edited by Kristoffer Gansing and Inga Luchs, 124–40. Amsterdam: Institute of Network Cultures, 2020.

Mantere, Eerik, and Sanna Raudaskoski. "The Sticky Media Device." In *Media, Family Interaction and the Digitalization of Childhood*, 135–54. Northampton, MA: Edward

Elgar, 2017. https://www.elgaronline.com/edcollchap/edcoll/9781785366666/9781785366666.00018.xml.
Marcelline, Marco. "Mastodon Sees Dip in Active Users After Twitter Exodus Surge." *PCMAG*, January 8, 2023. https://www.pcmag.com/news/mastodons-active-users-are-declining.
Marcuse, Herbert. "Repressive Tolerance." In *A Critique of Pure Tolerance*, edited by Robert P Wolff, Barrington Moore, and Herbert Marcuse, 81–123. Boston: Beacon Press, 1969.
Markham, Annette, and Elizabeth Buchanan. "Ethical Decision-Making and Internet Research: Recommendations from the Aoir Ethics Working Committee (Version 2.0)." Association of Internet Researchers, 2012. https://pure.au.dk/ws/files/55543125/aoirethics2.pdf.
Marrazzo, Vincent. "The Federalists of the Internet? What Online Platforms Can Learn from Reddit's Decentralized Content Moderation Scheme." *Nebraska Law Review*, January 19, 2023. https://lawreview.unl.edu/federalists-internet-what-online-platforms-can-learn-reddit%E2%80%99s-decentralized-content-moderation.
Martin, Aryn, Natasha Myers, and Ana Viseu. "The Politics of Care in Technoscience." *Social Studies of Science* 45, no. 5 (2015): 625–41.
Martin, Michel. "Elon Musk Calls Himself a Free Speech Absolutist. What Could Twitter Look Like Under His Leadership?" *NPR*, October 8, 2022, sec. Technology. https://www.npr.org/2022/10/08/1127689351/elon-musk-calls-himself-a-free-speech-absolutist-what-could-twitter-look-like-un.
Marwick, Alice E. "Morally Motivated Networked Harassment as Normative Reinforcement." *Social Media + Society* 7, no. 2 (April 1, 2021): 20563051211021378. https://doi.org/10.1177/20563051211021378.
Marwick, Alice E., and danah boyd. "I Tweet Honestly, I Tweet Passionately: Twitter Users, Context Collapse, and the Imagined Audience." *New Media & Society*, July 2010. https://doi.org/10.1177/1461444810365313.
Massanari, Adrienne. "#Gamergate and the Fappening: How Reddit's Algorithm, Governance, and Culture Support Toxic Technocultures." *New Media & Society* 19, no. 3 (March 1, 2017): 329–46. https://doi.org/10.1177/1461444815608807.
Massanari, Adrienne. *Gaming Democracy: How Silicon Valley Leveled Up the Far Right*. Cambridge, MA: MIT Press, 2024. https://mitpressbookstore.mit.edu/book/9780262549417.
Mastodon hosted on neuromatch.social. "Neuromatch Social." Accessed June 4, 2023. https://neuromatch.social/about.
"Mastodon.Social—Mastodon." February 24, 2017. https://web.archive.org/web/20170224052802/https://mastodon.social/about/more.
Matias, J. Nathan. "The Civic Labor of Volunteer Moderators Online." *Social Media + Society* 5, no. 2 (April 1, 2019): 2056305119836778. https://doi.org/10.1177/2056305119836778.
Mattern, Shannon. "Maintenance and Care." *Places Journal*, November 20, 2018. https://doi.org/10.22269/181120.
Maxwell, Richard, and Toby Miller. "Environmental Materialism and Media Globalization." In *The Routledge Handbook of Digital Media and Globalization*, 34–42. New York: Routledge, 2021.
McIlwain, Charlton D. *Black Software: The Internet and Racial Justice, from the AfroNet to Black Lives Matter*. New York: Oxford University Press, 2020.
Mehta, Ivan. "Google Says YouTube Shorts Has Crossed 50 Billion Daily Views." *TechCrunch* (blog), February 3, 2023. https://techcrunch.com/2023/02/03/google-says-youtube-shorts-has-crossed-50-billion-daily-views/.
Mejia, Robert. "A Pressure Chamber of Innovation: Google Fiber and Flexible Capital." *Communication and Critical/Cultural Studies* 12, no. 3 (July 3, 2015): 289–308. https://doi.org/10.1080/14791420.2015.1027240.
Mejias, Ulises A., and Nick Couldry. *The Costs of Connection: How Data Is Colonizing Human Life and Appropriating It for Capitalism*. Stanford, CA: Stanford University Press, 2019.

"Meta Reports Fourth Quarter and Full Year 2022 Results." February 1, 2023. https://investor.fb.com/investor-news/press-release-details/2023/Meta-Reports-Fourth-Quarter-and-Full-Year-2022-Results/default.aspx.

milhouse, life's coming up. "Don't Use Randi Harper's Blocklist. Use Naziblocker Instead." *Medium* (blog), January 7, 2017. https://medium.com/@obvious_humor/dont-use-randi-harper-s-blocklist-use-naziblocker-instead-db16cd666d49.

Mirov, Lev. "The Desert, Blooming." In *Sunvault: Stories of Solarpunk and Eco-Speculation*, edited by Phoebe Wagner and Brontë Christopher Wieland, 103–14. Nashville, TN: Upper Rubber Boot Books, 2017.

Miserandino, Christine. "The Spoon Theory." 2003. https://web.archive.org/web/20191117210039/https://butyoudontlooksick.com/articles/written-by-christine/the-spoon-theory/.

"Misfit Loser Zealot Club." Accessed January 11, 2024. https://misfitloserzealot.club/?ref=privacy.thenexus.today.

Mol, Annemarie, Ingunn Moser, and Jeannette Pols. "Care: Putting Practice into Theory." In *Care in Practice: On Tinkering in Clinics, Homes and Farms*, edited by Annemarie Mol, Ingunn Moser, and Jeannette Pols, 1st ed., 8:7–25. VerKörperungen/MatteRealities—Perspektiven Empirischer Wissenschaftsforschung. Bielefeld, Germany: transcript Verlag, 2010. https://doi.org/10.14361/transcript.9783839414477.

Molyneux, Logan, and Shannon C. McGregor. "Legitimating a Platform: Evidence of Journalists' Role in Transferring Authority to Twitter." *Information, Communication & Society* 25, no. 11 (August 18, 2022): 1577–95. https://doi.org/10.1080/1369118X.2021.1874037.

Monea, Alex. *The Digital Closet: How the Internet Became Straight*. Cambridge, MA: MIT Press, 2022.

Monge Roffarello, Alberto, and Luigi De Russis. "Towards Understanding the Dark Patterns That Steal Our Attention." In *Extended Abstracts of the 2022 CHI Conference on Human Factors in Computing Systems*, 1–7. CHI EA '22. New York: Association for Computing Machinery, 2022. https://doi.org/10.1145/3491101.3519829.

Monserrate, Steven Gonzalez. "The Staggering Ecological Impacts of Computation and the Cloud." *The MIT Press Reader* (blog), February 14, 2022. https://thereader.mitpress.mit.edu/the-staggering-ecological-impacts-of-computation-and-the-cloud/.

Moots, Glenn A. "The Covenant Tradition of Federalism: The Pioneering Studies of Daniel J. Elazar." In *The Ashgate Research Companion to Federalism*, edited by Ann Ward and Lee Ward, 391–412. Burlington, VT: Ashgate, 2009.

Morrison, Sara. "Meta Is Getting Data About You from Some Surprising Places." *Vox*, June 17, 2022. https://www.vox.com/recode/23172691/meta-tracking-privacy-hospitals.

Morstatter, Fred, Yunqiu Shao, Aram Galstyan, and Shanika Karunasekera. "From Alt-Right to Alt-Rechts: Twitter Analysis of the 2017 German Federal Election." In *Companion Proceedings of The Web Conference 2018*, 621–28. Republic and Canton of Geneva, CHE: International World Wide Web Conferences Steering Committee, 2018. https://doi.org/10.1145/3184558.3188733.

Mould, Oli, Jennifer Cole, Adam Badger, and Philip Brown. "Solidarity, Not Charity: Learning the Lessons of the COVID-19 Pandemic to Reconceptualise the Radicality of Mutual Aid." *Transactions of the Institute of British Geographers* 47, no. 4 (2022): 866–79. https://doi.org/10.1111/tran.12553.

Mozur, Paul. "A Genocide Incited on Facebook, With Posts from Myanmar's Military." *The New York Times*, October 15, 2018, sec. Technology. https://www.nytimes.com/2018/10/15/technology/myanmar-facebook-genocide.html.

Musambi, Evelyne. "Kenyan Facebook Moderators Accuse Meta of Not Negotiating Sincerely." *AP News*, October 16, 2023. https://apnews.com/article/kenya-facebook-content-moderators-lawsuit-1602c705457443d215e9530cd365f337.

"Musk Fires Outsourced Content Moderators Who Track Abuse on Twitter." November 14, 2022. https://www.cbsnews.com/news/elon-musk-twitter-layoffs-outsourced-content-moderators/.

Nafus, Dawn. "'Patches Don't Have Gender': What Is Not Open in Open Source Software." *New Media & Society* 14, no. 4 (June 1, 2012): 669–83. https://doi.org/10.1177/1461444811422887.

NBC News. "Google Buys YouTube for $1.65 Billion," October 9, 2006. https://www.nbcnews.com/id/wbna15196982.

nevillepark. "Code of Conduct · Issue #175 · Mastodon/Mastodon." *GitHub*, November 22, 2016. https://github.com/mastodon/mastodon/issues/175.

Newton, Casey. "Mastodon.Social Is an Open-Source Twitter Competitor That's Growing Like Crazy." *The Verge*, April 4, 2017. https://www.theverge.com/2017/4/4/15177856/mastodon-social-network-twitter-clone.

Newton, Casey. "Meta Is Building a Decentralized, Text-Based Social Network." March 10, 2023. https://www.platformer.news/p/meta-is-building-a-decentralized.

Nicholas, Josh. "Elon Musk Drove More Than a Million People to Mastodon—But Many Aren't Sticking Around." *The Guardian*, January 7, 2023, sec. News. https://www.theguardian.com/news/datablog/2023/jan/08/elon-musk-drove-more-than-a-million-people-to-mastodon-but-many-arent-sticking-around.

Nissenbaum, Helen. "Privacy as Contextual Integrity." *Washington Law Review* 79, no. 1 (February 1, 2004): 119.

Nix, Naomi. "Facebook Bans Hate Speech but Still Makes Money from White Supremacists." *Washington Post*, August 10, 2022. https://www.washingtonpost.com/technology/2022/08/10/facebook-white-supremacy-ads/.

Noddings, Nel. *Caring, a Feminine Approach to Ethics & Moral Education*. Berkeley: University of California Press, 1984.

NotJohnMastodon. "When You First . . . ." Reddit Comment. *R/Mastodon*, December 21, 2022. www.reddit.com/r/Mastodon/comments/zqfr4h/ama_with_eugen_rochko_founder_and_lead_developer/j1228og/.

O'Carroll, Lisa. "Disinformation Networks 'Flooded' X Before EU Elections, Report Says." *The Guardian*, July 12, 2024, sec. World news. https://www.theguardian.com/world/article/2024/jul/12/disinformation-networks-social-media-x-france-germany-italy-eu-elections.

Ojala, Maria, Ashlee Cunsolo, Charles A. Ogunbode, and Jacqueline Middleton. "Anxiety, Worry, and Grief in a Time of Environmental and Climate Crisis: A Narrative Review." *Annual Review of Environment and Resources* 46 (October 18, 2021): 35–58. https://doi.org/10.1146/annurev-environ-012220-022716.

Oliphant, The. "Oliphant.Social Mastodon Blocklists." *The Oliphant*, November 15, 2022. https://writer.oliphant.social/oliphant/the-oliphant-social-blocklist.

"OpenSocial Foundation Moving Standards Work to W3C Social Web Activity." December 16, 2014. https://www.w3.org/2014/12/opensocial.html.en.

Ordonez, Victor. "Key Takeaways from Facebook Whistleblower Frances Haugen's Senate Testimony." ABC News, October 5, 2021. https://abcnews.go.com/Politics/key-takeaways-facebook-whistleblower-frances-haugens-senate-testimony/story?id=80419357.

Ortega, Manuel. "GNU Social: Federation Against the Social Model of Twitter." *Las Indias in English* (blog), April 8, 2015. http://english.lasindias.com/gnu-social-federation-against-the-social-model-twitter.

Oxford English Dictionary. "Federation, n.," 2023. https://www.oed.com/dictionary/federation_n?tab=factsheet#4557410.

Pasek, Anne. "On Being Anxious About Digital Carbon Emissions." *Social Media + Society* 9, no. 2 (April 1, 2023): 20563051231177906. https://doi.org/10.1177/20563051231177906.

Pasek, Anne, Hunter Vaughan, and Nicole Starosielski. "The World Wide Web of Carbon: Toward a Relational Footprinting of Information and Communications Technology's Climate

Impacts." *Big Data & Society* 10, no. 1 (January 2023): 205395172311589. https://doi.org/10.1177/20539517231158994.

Patel, Nilay. "Can Mastodon Seize the Moment from Twitter?" *The Verge*, March 28, 2023. https://www.theverge.com/23658648/mastodon-ceo-twitter-interview-elon-musk-twitter.

Patringenaru, Ioana. "Who's Got Your Mail? Google and Microsoft, Mostly." *UC San Diego Today*, December 6, 2021. https://today.ucsd.edu/story/IMC2021_savage.

Paul, Kari. "We Tried Threads, Meta's New Twitter Rival. Here's What Happened." *The Guardian*, July 6, 2023, sec. Technology. https://www.theguardian.com/technology/2023/jul/06/we-tried-threads-metas-new-twitter-rival-heres-what-happened.

Paul, Kari, Lois Beckett, and Josh Taylor. "Twitter Suspends Accounts of Several Journalists Who Had Reported on Elon Musk." *The Guardian*, December 16, 2022, sec. Technology. https://www.theguardian.com/technology/2022/dec/15/twitter-suspends-accounts-journalists-musk.

Paul, Pamela. "Is Threads the Good Place?" *The New York Times*, March 28, 2024, sec. Opinion. https://www.nytimes.com/2024/03/28/opinion/threads-x-twitter-social-media.html.

Paulson, Susan. "Economic Growth Will Continue to Provoke Climate Change." *The Economist*, September 12, 2022. https://impact.economist.com/sustainability/circular-economies/economic-growth-will-continue-to-provoke-climate-change.

*Pepper & Carrot*. Pepper & Carrot/David Revoy. Hamburg, Germany: POPCOM, 2017.

Perez, Sarah. "Mastodon's Latest Release Makes the Open Source Twitter Alternative Easier to Use." *TechCrunch*, September 21, 2023. https://techcrunch.com/2023/09/21/mastodons-latest-release-makes-the-open-source-twitter-alternative-easier-to-use/.

Perez, Sarah. "Tumblr's 'Fediverse' Integration Is Still Being Worked on, Says Owner and Automattic CEO Matt Mullenweg." *TechCrunch* (blog), December 11, 2023. https://techcrunch.com/2023/12/11/tumblrs-fediverse-integration-is-still-being-worked-on-says-owner-and-automattic-ceo-matt-mullenweg/.

Peters, Jay. "The Hunt for the Next Twitter: All the News About Alternative Social Media Platforms." *The Verge*, April 17, 2023. https://www.theverge.com/23686584/twitter-alternative-social-media-platforms-mastodon-bluesky-activitypub-protocol.

Peters, Jay. "Twitter Is Blocking Links to Mastodon." *The Verge*, December 16, 2022. https://www.theverge.com/2022/12/15/23512113/twitter-blocking-mastodon-links-elon-musk-elonjet.

Petre, Caitlin, Brooke Erin Duffy, and Emily Hund. "'Gaming the System': Platform Paternalism and the Politics of Algorithmic Visibility." *Social Media + Society* 5, no. 4 (October 1, 2019): 2056305119879995. https://doi.org/10.1177/2056305119879995.

pierat. "Meta/Facebook Is Inviting Fediverse Admins Under NDA for 'Meetings.'" *Hacker News*, June 18, 2023. https://news.ycombinator.com/item?id=36384207.

Pierce, David. "Can ActivityPub Save the Internet?" *The Verge*, April 20, 2023. https://www.theverge.com/2023/4/20/23689570/activitypub-protocol-standard-social-network.

Pincus, Jon. "Fork It! It's Time for a Mastodon Hard Fork (UPDATED)." *The Nexus of Privacy*, April 28, 2024. https://privacy.thenexus.today/mastodon-hard-fork/.

Pixelfed Documentation. "API," October 17, 2022. https://docs.pixelfed.org/technical-documentation/api/.

Poell, Thomas. "Three Challenges for Media Studies in the Age of Platforms." *Television & New Media* 21, no. 6 (September 1, 2020): 650–57. https://doi.org/10.1177/1527476420918833.

Poo-Caamaño, Germán, Eric Knauss, Leif Singer, and Daniel M. German. "Herding Cats in a FOSS Ecosystem: A Tale of Communication and Coordination for Release Management." *Journal of Internet Services and Applications* 8, no. 1 (August 30, 2017): 12. https://doi.org/10.1186/s13174-017-0063-2.

Porter, Jon. "Instagram's Threads Surpasses 100 Million Users." *The Verge*, July 10, 2023. https://www.theverge.com/2023/7/10/23787453/meta-instagram-threads-100-million-users-milestone.

Porter, Jon. "Threads Launches for Nearly Half a Billion More Users in Europe." *The Verge*, December 14, 2023. https://www.theverge.com/2023/12/14/23953986/threads-european-union-launch-eu-meta-twitter-rival.

Potter, Martin. "Bad Actors Never Sleep: Content Manipulation on Reddit." In *The Dark Social*, edited by Toija Cinque, Alexia Maddox, and Robert W. Gehl, 46–58. New York: Routledge, 2024.

"The Problem with Personal Block Lists: A Blog." April 29, 2016. https://web.archive.org/web/20160429173205/https://sjwomble.wordpress.com/2016/04/28/the-problem-with-personal-block-lists/.

Prodromou, Evan. "Big Fedi, Small Fedi." *Evan Prodromou's Blog* (blog), December 27, 2023. https://evanp.me/2023/12/26/big-fedi-small-fedi/.

Prodromou, Evan. "This Is My First Post." Microblog. Identi.ca, May 18, 2008. https://web.archive.org/web/20130615111045/http://identi.ca/notice/1.

Pulliam, Roland X. "Federation Safety Enhancement Project (FSEP)," August 10, 2023. https://nivenly.org/docs/papers/fsep/.

Quirk, Kev. "Threads and the Fediverse." *Kev Quirk*, December 16, 2023. https://kevquirk.com/threads-and-the-fediverse.

Raghuram, Parvati. "Race and Feminist Care Ethics: Intersectionality as Method." *Gender, Place & Culture* 26, no. 5 (May 4, 2019): 613–37. https://doi.org/10.1080/0966369X.2019.1567471.

Raman, Aravindh, Sagar Joglekar, Emiliano De Cristofaro, Nishanth Sastry, and Gareth Tyson. "Challenges in the Decentralised Web: The Mastodon Case." In *Proceedings of the Internet Measurement Conference*, 217–29. IMC '19. Amsterdam, Netherlands: Association for Computing Machinery, 2019. https://doi.org/10.1145/3355369.3355572.

Ratta, Donatella Della. "Digital Socialism Beyond the Digital Social: Confronting Communicative Capitalism with Ethics of Care." *tripleC: Communication, Capitalism & Critique. Open Access Journal for a Global Sustainable Information Society* 18, no. 1 (January 13, 2020): 101–15. https://doi.org/10.31269/triplec.v18i1.1145.

Ratto, Matt, and Megan Boler, eds. *DIY Citizenship: Critical Making and Social Media*. Cambridge, MA: The MIT Press, 2014.

Ray, Siladitya. "Meta Is Creating a Decentralized Twitter Alternative Reportedly Codenamed 'P92.'" *Forbes*, March 10, 2023. https://www.forbes.com/sites/siladityaray/2023/03/10/meta-is-working-on-a-decentralized-twitter-alternative-reportedly-codenamed-p92/.

Raymond, Eric S. *The Cathedral and the Bazaar: Musings on Linux and Open Source by an Accidental Revolutionary*. Beijing: O'Reilly, 2001.

Reagle, Joseph. "Naive Meritocracy and the Meanings of Myth." *Ada: A Journal of Gender, New Media, and Technology*, no. 11 (May 2017). https://scholarsbank.uoregon.edu/xmlui/handle/1794/26770.

Reagle, Joseph M. *Good Faith Collaboration: The Culture of Wikipedia*. History and Foundations of Information Science. Cambridge, MA: MIT Press, 2010.

Reina-Rozo, Juan David. "Art, Energy and Technology: The Solarpunk Movement." *International Journal of Engineering, Social Justice, and Peace* 8, no. 1 (March 5, 2021): 47–60. https://doi.org/10.24908/ijesjp.v8i1.14292.

Renzi, Alessandra. *Hacked Transmissions: Technology and Connective Activism in Italy*. Minneapolis: University of Minnesota Press, 2020.

"REST API Quick Start | PeerTube Documentation." Accessed October 13, 2023. https://docs.joinpeertube.org/api/rest-getting-started.

Reuters. "Trump's New App Tops Downloads in Apple Store." NBC News, February 21, 2022. https://www.nbcnews.com/politics/donald-trump/trump-s-new-app-tops-downloads-apple-store-n1289441.

"Revealed: Trump Campaign Strategy to Deter Millions of Black Americans from Voting in 2016." 2020. https://www.youtube.com/watch?v=KIf5ELaOjOk.

Revoy, David. “Don’t Roll out the Red Carpet for Them.” *David Revoy*, December 14, 2023. https://www.davidrevoy.com/article1009/dont-roll-out-the-red-carpet-for-them.

Revoy, David. “License F.A.Q.” *David Revoy*, 2024. https://www.davidrevoy.com/static8/license-faq.

Revoy, David. “Pepper & Carrot (MOTION COMIC) Episode 6: The Potion Contest.” *Touhoppai Tube*, March 25, 2018. https://peertube.touhoppai.moe/w/g4Kni4fEskBSdurcY8K9AM.

Revoy, David. “The Derivative Pepper&Carrot Motion Comic Project by Nikolai Mamashev—David Revoy.” Blog. *David Revoy*, September 1, 2016. https://www.davidrevoy.com/article580/motion-comic-project-by-nikolai-mamashev.

Rhodes, Margaret. “Like Twitter but Hate the Trolls? Try Mastodon.” *Wired*, April 13, 2017. https://www.wired.com/2017/04/like-twitter-hate-trolls-try-mastodon/.

Rixen, Jan Ole, Luca-Maxim Meinhardt, Michael Glöckler, Marius-Lukas Ziegenbein, Anna Schlothauer, Mark Colley, Enrico Rukzio, and Jan Gugenheimer. “The Loop and Reasons to Break It: Investigating Infinite Scrolling Behaviour in Social Media Applications and Reasons to Stop.” *Proceedings of the ACM Human-Computer Interaction* 7, no. MHCI (September 13, 2023): 228:1–228:22. https://doi.org/10.1145/3604275.

Ro. “Dev Diaries 02 & The Numbers for November.” *Patreon*, November 30, 2017. https://www.patreon.com/posts/dev-diaries-02-15625928.

Ro. “Dev Diary—Mini Update 02 - BIG CHANGES.” *Patreon*, May 8, 2018. https://www.patreon.com/posts/dev-diary-mini-18682166.

Ro. “Recuerda PV : Chapter 1—The Start, Part 1.” Blog. roiskinda.cool. November 24, 2022. https://roiskinda.cool/2022/11/recuerda-pv-chapter-1-the-start-part-1.html.

Ro. “Recuerda PV : Chapter 2—The Start, Part 2.” Blog. roiskinda.cool, December 2, 2022. https://roiskinda.cool/2022/12/recuerda-pv-chapter-2-the-start-part-2.html.

Ro. “Recuerda PV : Chapter 3—The Name.” Blog. roiskinda.cool, December 9, 2022. https://roiskinda.cool/2022/12/recuerda-pv-chapter-3-the-name.html.

Ro. “The Big Fat Intro.” *Vimeo*, July 16, 2018. https://vimeo.com/user52279813.

Roberts, Sarah T. *Behind the Screen: The Hidden Digital Labor of Commercial Content Moderation*. Urbana-Champaign: University of Illinois, 2014.

Robertson, Adi. “OLPC’s $100 Laptop Was Going to Change the World—Then It All Went Wrong.” *The Verge*, April 16, 2018. https://www.theverge.com/2018/4/16/17233946/olpcs-100-laptop-education-where-is-it-now.

Robertson, Adi. “With Linux’s Founder Stepping Back, Will the Community Change Its Culture?” *The Verge*, September 21, 2018. https://www.theverge.com/2018/9/21/17883442/linux-founder-linus-torvalds-apology-code-of-conduct-change-enforcement.

Rochko, Eugen. “I Don’t Think Shared Server Blocklists Are a Good Idea.” *Mastodon*, August 30, 2018. https://mastodon.social/@Gargron/100640826979121551.

Rochko, Eugen. “Learning from Twitter’s Mistakes.” *Eugen Rochko* (blog), March 3, 2017. https://medium.com/@Gargron/learning-from-twitters-mistakes-c272d67bba76.

Rochko, Eugen. “Mastodon and the W3C.” *Hacker Noon*, September 10, 2017. https://hackernoon.com/mastodon-and-the-w3c-f75f376f422.

Rochko, Eugen. “Mastodon Forms New U.S. Non-Profit.” *Mastodon Blog*, April 27, 2024. https://blog.joinmastodon.org/2024/04/mastodon-forms-new-u.s.-non-profit/.

Rochko, Eugen. “Show HN: A New Decentralized Microblogging Platform.” *Hacker News*, October 5, 2016. https://news.ycombinator.com/item?id=12646083.

Rochko, Eugen. “What to Know About Threads.” *Mastodon Blog*, July 5, 2023. https://blog.joinmastodon.org/2023/07/what-to-know-about-threads/.

Rochko, Eugen. “Write a Terms of Service and Privacy Policy · Issue #115 · Mastodon/Mastodon.” *GitHub*, November 1, 2016. https://github.com/mastodon/mastodon/issues/115.

Rochko, Eugen (@Gargron@mastodon.social). “My Wife Is Looking Forward to Deleting Her Instagram Account Once She Can Connect with the Same Folks from Her Mastodon

Account. Being Able. . . ." Mastodon post. *Mastodon*, December 16, 2023. https://mastodon.social/@Gargron/111591822065763831.

Rodríguez, Clemencia. "Citizens' Media." In *The International Encyclopedia of Communication*, edited by Wolfgang Donsbach, 1–3. Chichester, UK: John Wiley & Sons, 2008. https://doi.org/10.1002/9781405186407.wbiecc029.

Rodríguez, Clemencia. "Studying Media at the Margins: Learning from the Field." In *Media Activism in the Digital Age*, 49–61. London: Routledge, 2017. https://doi.org/10.4324/9781315393940-5.

Rodríguez, Clemencia. "The Bishop and His Star: Citizens' Communication in Southern Chile." In *Contesting Media Power: Alternative Media in a Networked World*, edited by Nick Couldry and James Curran, 177–94. Oxford: Rowman & Littlefield, 2003.

Rodríguez, Clemencia, Benjamin Ferron, and Kristin Shamas. "Four Challenges in the Field of Alternative, Radical and Citizens' Media Research." *Media, Culture & Society* 36, no. 2 (March 1, 2014): 150–66. https://doi.org/10.1177/0163443714523877.

Roscam Abbing, Roel. "'This Is a Solar-Powered Website, Which Means It Sometimes Goes Offline': A Design Inquiry into Degrowth and ICT." *LIMITS Workshop on Computing Within Limits*, June 14, 2021. https://doi.org/10.21428/bf6fb269.e78d19f6.

Rose, Arthur. "Mining Memories with Donald Trump in the Anthropocene." *Modern Fiction Studies* 64, no. 4 (2018): 701–22.

Roth, Yoel, and Samantha Lai. "Securing Federated Platforms: Collective Risks and Responses." *Journal of Online Trust and Safety* 2, no. 2 (February 28, 2024). https://doi.org/10.54501/jots.v2i2.171.

Rubin, Herbert J., and Irene Rubin. *Qualitative Interviewing: The Art of Hearing Data*. 3rd ed. Thousand Oaks, CA: Sage, 2012.

Russell, Andrew L., James L. Pelkey, and Loring Robbins. "The Business of Internetworking: Standards, Start-Ups, and Network Effects." *Business History Review* 96, no. 1 (March 2022): 109–44. https://doi.org/10.1017/S000768052100074X.

Sander-Staudt, Mary. "Care Ethics." *Internet Encyclopedia of Philosophy* (blog). Accessed January 30, 2024. https://iep.utm.edu/care-ethics/.

Sayedi, Amin. "Real-Time Bidding in Online Display Advertising." *Marketing Science* 37, no. 4 (August 2018): 553–68. https://doi.org/10.1287/mksc.2017.1083.

Schneider, Nathan. "Admins, Mods, and Benevolent Dictators for Life: The Implicit Feudalism of Online Communities." *New Media & Society* 24, no. 9 (September 1, 2022): 1965–85. https://doi.org/10.1177/1461444820986553.

Schneider, Nathan. *Governable Spaces: Democratic Design for Online Life*. Oakland: University of California Press, 2024.

Seirdy. "De-Federating P92." *Sierdy*, June 20, 2023. https://seirdy.one/posts/2023/06/20/defederating-p92/.

Sevignani, Sebastian. "Facebook vs. Diaspora: A Critical Study." In *Unlike Us Reader: Social Media Monopolies and Their Alternatives*, edited by Geert Lovink and Miriam Rausch, 325–37. Amsterdam: Institute of Network Cultures, 2013.

Siddik, Md Abu Bakar, Arman Shehabi, and Landon Marston. "The Environmental Footprint of Data Centers in the United States." *Environmental Research Letters* 16, no. 6 (May 2021): 064017. https://doi.org/10.1088/1748-9326/abfba1.

Siele, Martin K. N. "'It's Been Tough for Us': Meta's Kenyan Content Moderators Say They'll Keep Fighting." *Rest of World*, May 22, 2023. https://restofworld.org/2023/meta-content-moderators-kenya-fired-unionize/.

Sinnreich, Aram. *Mashed up: Music, Technology, and the Rise of Configurable Culture*. Science/Technology/Culture. Amherst: University of Massachusetts Press, 2010.

Smythe, Dallas. "On the Audience Commodity and Its Work." In *Dependency Road: Capitalism, Consciousness, and Canada*, 253–79. Norwood, NJ: Ablex, 1981.

Social.coop. "How to Make the Fediverse Your Own," April 15, 2024. https://wiki.social.coop/wiki/How_to_Make_the_Fediverse_Your_Own#Implementing_a_code_of_conduct.

The Social Media Alternatives Project. "Mastodon | Awoo.Space | About Text." June 21, 2017. https://www.socialmediaalternatives.org/archive/items/show/576.

"Socialwg/2015-12-02-Minutes—W3C Wiki," December 2, 2015. https://www.w3.org/wiki/Socialwg/2015-12-02-minutes.

"Socialwg/2016-06-07-Minutes—W3C Wiki," June 7, 2016. https://www.w3.org/wiki/Socialwg/2016-06-07-minutes.

"Socialwg/2016-11-17-Minutes—W3C Wiki," November 17, 2016. https://www.w3.org/wiki/Socialwg/2016-11-17-minutes.

"Socialwg/2017-12-05-Minutes—W3C Wiki," December 5, 2017. https://www.w3.org/wiki/Socialwg/2017-12-05-minutes.

Spade, Dean. "Solidarity Not Charity: Mutual Aid for Mobilization and Survival." *Social Text* 38, no. 1 (March 1, 2020): 131–51. https://doi.org/10.1215/01642472-7971139.

Stark, Luke. "Algorithmic Psychometrics and the Scalable Subject." *Social Studies of Science* 48, no. 2 (April 2018): 204–31. https://doi.org/10.1177/0306312718772094.

Steele, Catherine Knight. *Digital Black Feminism*. New York: NYU Press, 2021.

Stevenson, Michael, and Carolina Valente Pinto. "Distinction and Alternative Tech: Exploring the Techno-Critical Disposition." *New Media & Society*, March 28, 2024, 14614448241239579. https://doi.org/10.1177/14614448241239579.

Stiegler, Bernard. *For a New Critique of Political Economy*. Cambridge: Polity, 2010.

Stocking, Galen, Amy Mitchell, Katerina Eva Matsa, Regina Widjaya, Mark Jurkowitz, Shreenita Ghosh, Aaron Smith, Sarah Naseer, and Christopher St Aubin. "The Role of Alternative Social Media in the News and Information Environment." *Pew Research Center's Journalism Project* (blog), October 6, 2022. https://www.pewresearch.org/journalism/2022/10/06/the-role-of-alternative-social-media-in-the-news-and-information-environment/.

Stray, Jonathan, Ravi Iyer, and Helen Puig Larrauri. "The Algorithmic Management of Polarization and Violence on Social Media." Knight First Amendment Institute, August 22, 2023. http://knightcolumbia.org/content/the-algorithmic-management-of-polarization-and-violence-on-social-media.

Striphas, Ted. "Algorithmic Culture." *European Journal of Cultural Studies* 18, no. 4–5 (August 1, 2015): 395–412. https://doi.org/10.1177/1367549415577392.

strugee. "ActivityPub Support · Issue #1557 · Tootsuite/Mastodon." *GitHub*, April 11, 2017. https://github.com/tootsuite/mastodon/issues/1557.

Stubblebine, Tony. "Medium Embraces Mastodon." *The Medium Blog* (blog), January 12, 2023. https://blog.medium.com/medium-embraces-mastodon-19dcb873eb11.

Sugimoto, Cassidy, Carolyn Hank, Timothy Bowman, and Jeffrey Pomerantz. "Friend or Faculty: Social Networking Sites, Dual Relationships, and Context Collapse in Higher Education." *First Monday* 20, no. 3 (February 27, 2015). https://doi.org/10.5210/fm.v20i3.5387.

tedu. "ActivityPub as It Has Been Understood." *flak*, May 28, 2021. https://flak.tedunangst.com/post/ActivityPub-as-it-has-been-understood.

Terranova, Tiziana. "Free Labor: Producing Culture for the Digital Economy." *Social Text* 18, no. 2 (2000): 33–58.

Thorbecke, Catherine. "Layoffs, Ultimatums, and an Ongoing Saga over Blue Check Marks: Elon Musk's First Month at Twitter." CNN, November 27, 2022. https://www.cnn.com/2022/11/27/tech/elon-musk-one-month-twitter/index.html.

Tiara, Creatrix. "Mastodon 101: A Queer-Friendly Social Network You're Gonna Like a Lot." *Autostraddle* (blog), October 11, 2017. https://www.autostraddle.com/mastodon-101-a-queer-friendly-social-network-youre-gonna-like-a-lot-390948/.

Tilley, Sean. "One Mammoth of a Job: An Interview with Eugen Rochko of Mastodon." *We Distribute* (blog), July 9, 2018. https://medium.com/we-distribute/one-mammoth-of-a-job-an-interview-with-eugen-rochko-of-mastodon-23b159d6796a.

Tkacz, Nathaniel. “The Politics of Forking Paths.” In *Critical Point of View a Wikipedia Reader*, edited by Geert Lovink and Nathaniel Tkacz, 94–107. Amsterdam: Institute of Network Cultures, 2011.

Torba, Andrew. “Responding to Rachel Maddow’s Attack on Gab.” Gab News, July 26, 2022. https://news.gab.com/2022/07/responding-to-rachel-maddows-attack-on-gab/.

Torba, Andrew. “The Path Forward: Build and Balkanize.” Gab News, November 10, 2022. https://news.gab.com/2022/11/the-path-forward-build-and-balkanize/.

Torba, Andrew. “There Is No Freedom of Speech Without Freedom of Reach.” Gab News, November 25, 2022. https://news.gab.com/2022/11/there-is-no-freedom-of-speech-without-freedom-of-reach/.

Torba, Andrew. “Your Duty to Wake Up Those Around You.” Gab News, May 23, 2022. https://news.gab.com/2022/05/your-duty-to-wake-up-those-around-you/.

Treisman, Rachel. “Got a Question for Twitter’s Press Team? The Answer Will Be a Poop Emoji.” *NPR*, March 20, 2023, sec. Business. https://www.npr.org/2023/03/20/1164654551/twitter-poop-emoji-elon-musk.

Trev. “GNU Social, Pleroma, and the Mastodon Culture Conflict.” *Medium* (blog), June 9, 2018. https://medium.com/@Trevatos/gnu-social-pleroma-and-the-mastodon-culture-conflict-3b10872937c2.

Trottier, Chris. “Federation with #Meta Actually Hurts Meta.” Microblog. Atomic Poet, June 28, 2023. https://atomicpoet.org/notice/AX9zOBSSW6gg06h9t2.

Valens, Ana. “Mastodon Is Crumbling—And Many Blame Its Creator.” *The Daily Dot*, January 18, 2019. https://www.dailydot.com/debug/mastodon-fediverse-eugen-rochko/.

Vezina, Adam. “Defederate Meta.” Cache Rules, June 21, 2023. https://www.cacherules.com/blog/2023/6/defederate-meta/?ref=privacy.thenexus.today.

Vossen, Bas van der, and Billy Christmas. “Libertarianism.” In *The Stanford Encyclopedia of Philosophy*, edited by Edward N. Zalta and Uri Nodelman, Fall 2023. Metaphysics Research Lab, Stanford University, 2023. https://plato.stanford.edu/archives/fall2023/entries/libertarianism/.

W3C. “Social Web Working Group—Publications.” January 23, 2018. https://www.w3.org/groups/wg/social/publications/.

W3C. “Software and Document License—2015 Version.” 2015. https://www.w3.org/copyright/software-license-2015/.

W3C.org. “Social Web Working Group Charter.” July 21, 2014. https://www.w3.org/2013/socialweb/social-wg-charter.html.

Wagner, Kurt. “Musk’s X 2023 Ad Sales Projected to Slump to About $2.5 Billion.” *Bloomberg.Com*, December 12, 2023. https://www.bloomberg.com/news/articles/2023-12-12/musk-s-x-2023-ad-sales-projected-to-slump-to-about-2-5-billion.

Wagner, Kurt. “This Is How Facebook Collects Data on You Even If You Don’t Have an Account.” *Vox*, April 20, 2018. https://www.vox.com/2018/4/20/17254312/facebook-shadow-profiles-data-collection-non-users-mark-zuckerberg.

Wagner, Phoebe, and Brontë Christopher Wieland, eds. *Sunvault: Stories of Solarpunk and Eco-Speculation*. Nashville, TN: Upper Rubber Boot Books, 2017.

Wähner, Marco, Annika Deubel, Johannes Breuer, and Katrin Weller. “ ‘Don’t Research Us’—How Mastodon Instance Rules Connect to Research Ethics.” *Publizistik*, August 12, 2024. https://doi.org/10.1007/s11616-024-00855-6.

Wajcman, Judy. *Pressed for Time: The Acceleration of Life in Digital Capitalism*. Chicago: University of Chicago Press, 2015. http://site.ebrary.com/lib/qut/docDetail.action?docID=10967678.

Waldman, Ari, and Kirsten Martin. “Governing Algorithmic Decisions: The Role of Decision Importance and Governance on Perceived Legitimacy of Algorithmic Decisions.” *Big Data & Society* 9, no. 1 (January 1, 2022): 20539517221100449. https://doi.org/10.1177/20539517221100449.

Watson, T. X. "The Boston Hearth Project." In *Sunvault: Stories of Solarpunk and Eco-Speculation*, edited by Phoebe Wagner and Brontë Christopher Wieland, 14–25. Nashville, TN: Upper Rubber Boot Books, 2017.

Wessalowski, Nate, and Mara Karagianni. "From Feminist Servers to Feminist Federation." *Minor Tech* 12, no. 1 (2023): 192–208.

White, Richard J., and Colin C. Williams. "Anarchist Economic Practices in a 'Capitalist' Society: Some Implications for Organisation and the Future of Work." *Ephemera: Theory & Politics in Organization* 14, no. 4 (2014): 947–71.

Wong, Julia Carrie. "Banning Trump Won't Fix Social Media: 10 Ideas to Rebuild Our Broken Internet—by Experts." *The Guardian*, January 16, 2021, sec. Media. https://www.theguardian.com/media/2021/jan/16/how-to-fix-social-media-trump-ban-free-speech.

Wylie, Christopher. *Mindf*ck: Cambridge Analytica and the Plot to Break America*. New York: Random House, 2019.

X. "X Terms of Service," 2024. https://x.com/en/tos.

X, Marcia. "#Fediblock, a Tiny History." Artist Marcia X, November 5, 2024. https://subscriptions.boricua.style/fediblock-a-tiny-history-2/.

X, Marcia, and Ra'il I'Nasah Kiam. "Blackness in the Fediverse: An Interview with Artist Marcia X." *Logic*, February 2024.

Yaszek, Lisa. "Afrofuturism, Science Fiction, and the History of the Future." *Socialism and Democracy* 20, no. 3 (November 1, 2006): 41–60. https://doi.org/10.1080/08854300600950236.

Young, Iris Marion. "Hybrid Democracy: Iroquois Federalism and the Postcolonial Project." In *Political Theory and the Rights of Indigenous Peoples*, edited by Duncan Ivison, Paul Patton, and Will Sanders, 237–58. Cambridge: Cambridge University Press, 2000.

Zuboff, Shoshana. *The Age of Surveillance Capitalism: The Fight for a Human Future at the New Frontier of Power*. 1st ed. New York: PublicAffairs, 2019.

Zuckerman, Ethan. "How Social Media Could Teach Us to Be Better Citizens." *Journal of E-Learning and Knowledge Society* 18, no. 3 (December 28, 2022): 36–41. https://doi.org/10.20368/1971-8829/1135818.

Zulli, Diana. "Capitalizing on the Look: Insights into the Glance, Attention Economy, and Instagram." *Critical Studies in Media Communication* 35, no. 2 (March 15, 2018): 137–50. https://doi.org/10.1080/15295036.2017.1394582.

Zulli, Diana, Miao Liu, and Robert Gehl. "Rethinking the 'Social' in 'Social Media': Insights into Topology, Abstraction, and Scale on the Mastodon Social Network:" *New Media & Society*, July 22, 2020. https://doi.org/10.1177/1461444820912533.

# Index

*For the benefit of digital users, indexed terms that span two pages (e.g., 52–53) may, on occasion, appear on only one of those pages.*